MATARIKI

MATARIKI
The Star of the Year

RANGI MATAMUA

First published in 2017 by Huia Publishers
39 Pipitea Street, PO Box 12280
Wellington, Aotearoa New Zealand
www.huia.co.nz

Reprinted in 2018, 2019, 2020, 2021

ISBN 978-1-77550-325-5

Cover image: Te Haunui Tuna
Back cover photograph of author © University of Waikato

A catalogue record for this book is available from the National Library of New Zealand.

Photograph: Erica Sinclair

Matariki and other stars, as seen from the porch of
Te Whai a te Motu meeting house, Mataatua Marae, Ruatāhuna.
Photograph: Te Rawhitiroa Bosch 2015.

Ki taku koroua ki a Timi

Tērā ko Matariki e ārau mai ana
E hahu ake nei i ōku maharatanga mōu

To my grandfather Timi

There I see Matariki gathering above
Recalling to mind my memories of you

CONTENTS

ACKNOWLEDGEMENTS

Mānawa maiea te putanga o Matariki
Mānawa maiea te ariki o te rangi
Mānawa maiea te mātahi o te tau[1]

Hail the rise of Matariki
Hail the lord of the sky
Hail the New Year[2]

I am deeply indebted to a number of organisations and individuals for their support in the compilation of this book. To all the people who gave me their time and openly shared their knowledge, I will be forever grateful. I must acknowledge the support I have received from staff of the University of Waikato, and in particular the Faculty of Māori and Indigenous Studies and the University of Waikato Library. Special note must go to my colleagues and friends Hemi Whaanga and Hōhepa Tuahine. Your efforts have been vital in the writing of this book. Thank you Te Taura Whiri i te reo Māori, the Royal Society of New Zealand, Fulbright New Zealand, Ngā Pae o te Māramatanga and the Society for Māori Astronomy and Traditions (SMART) for the role you have all played in this publication. To my many friends who have been part of this journey, those who have attended my presentations, asked questions, given advice and shared stories, I thank you. I am especially grateful to my companions from Te Mata Pūnenga and Te Panekiretanga o te reo Māori, and to our mentors Wharehuia Milroy, Timoti Karetu and Pou Temara. My uncle Pou has been my mentor and confidant for many years, and I will forever be indebted for his support and guidance. To my relations from Te Urewera, and to my whānau, for all your love throughout the years I say thank you. Finally, to the great stars in my life, my wife Marley and our children, for all your patience, understanding, support and unwavering commitment to me and this undertaking. This is my gift to you.

Rangi

MATARIKI – TE WHETŪ TAPU O TE TAU[1]
MATARIKI – THE STAR OF THE YEAR

INTRODUCTION

Astronomy, the study of the cosmos, is an endeavour that unites all peoples throughout the world. Since early times mankind has looked into the heavens seeking knowledge, understanding and inspiration. The many celestial objects above the earth were viewed as gods, as ancestors and supernatural beings, and their stories were embroidered into the tapestry of the night sky. Over time, astronomical observations became important localised knowledge, and the lights of the night sky emerged as markers, signalling the change of the seasons, the migration and spawning of various species and the many cycles of the environment.

For the ancestors of the Māori, astronomy was interwoven into all facets of life. Meticulous observations of the movements of the stars and the planets, the changing position of the sun, the phases of the moon and the appearance of anomalies such as comets and meteors were recorded and handed down from generation to generation as part of Māori oral tradition. Māori astronomy experts, tohunga kōkōrangi would observe the sky from sunset to sunrise, noting the appearance of the stars, and deducing meanings from their position, colour, movement and brightness. This detailed knowledge was taught to a select few via the whare kōkōrangi, or the Māori astronomical house of learning. It was within the confines of these institutions that students were taught the names of hundreds of stars and constellations, their meanings, their signs and their connection to the activities of man on the earth.

At times, the lights of the sky were used in a very pragmatic sense, and the appearance of certain stars on the horizon before dawn would inform the community that a new season had begun, or that it was time to harvest a particular food item. On other occasions the appearances of various celestial bodies were intertwined with Māori spirituality and beliefs. For instance, a planet near the crescent moon might indicate that a war party was about to attack, or a large lunar halo would warn the people that an important person was near death. Māori derived all kind of omen, message and meaning from the heavens, and believed the stars foretold their fortune and future.[2] In this way Māori star lore was, and still remains, a blending together of both astronomy and astrology, and while there is undoubtedly robust science within the Māori study of the night sky, the spiritual component has always been of equal importance.

Astronomy was central in the populating of the islands of the Pacific. Polynesian explorers used the stars, among other techniques, to navigate their double-hulled canoes across the greatest expanse of water on the planet, to settle many islands including Aotearoa. Māori used the stars to understand and interact with their environment, looking for the rising and setting of different points of light to tell them when to plant, when to harvest, when to hunt, when to fish, when to build, when to travel, when to celebrate and when to pray. Furthermore, this knowledge was embedded into the culture, the language and even the landscape. Evidence of Māori astronomy is recorded in numerous traditional Māori songs, proverbs, idioms and incantations. A number of landmarks, buildings and individuals bear the names of the objects of the sky, and the 'Whānau Mārama',

The island of Matariki near Riverton, South Island.
Photograph: Erica Sinclair 2015.

the Māori term for the objects that inhabit the sky, including the sun, moon, planets, stars and other forms of light, were at times worshipped as deities connected to both the sky and the earth.

When the first European settlers arrived to Aotearoa in the late 1700s and early 1800s, they were astounded by the large amount of astronomical knowledge maintained by Māori. It was widely recognised that Māori knew much more about the night sky than their European counterparts, and their ability to observe distant nebular and other objects with their naked eye was unequalled.[3] However, colonisation and its many attributes infiltrated to the core of Māori society, affecting all cultural practices, including Māori astronomy. By the start of the twentieth century, many Māori customs were no longer practised, and only a few individuals with detailed knowledge of Māori astronomy remained. A number of Pākehā ethnologists and historians wrote as much as they could about Māori star lore. In particular, Elsdon Best compiled the book *Astronomical Knowledge of the Maori* and also wrote a large section on stars within his tribal ethnography, *Tuhoe*.[4] However, even he described this effort as unsatisfactory and incomplete, and the remaining pieces of Māori astronomy that were scattered across various books and articles were mourned by Best as fragments of a lost practice:

> The available data concerning Maori sky-lore is now exhausted, and this account must be closed. The knowledge gained by us of this subject is meagre and unsatisfactory, but it is now too late to remedy the deficiency.[5]

Fortunately, Best's comments are incorrect, and this book itself is built upon a 400-page manuscript that deals exclusively with Māori star lore. In the late 1800s, Te Kōkau and his son Rāwiri Te Kōkau of Ruatāhuna were interviewed by Elsdon Best as part of his research for the publication *Astronomical Knowledge of the Maori*. At the completion of these interviews, Best gave to his informants a star map. This map was used by Te Kōkau and his son to compile a manuscript, which they began together in 1898 and Rawiri completed in 1933. On his deathbed, Rāwiri Te Kōkau handed this book to his grandson Timi Rāwiri, and in 1995 this book was then gifted to the grandson of Timi Rāwiri, the author of this book, *Matariki: The Star of the Year*. Therefore, after years of studying this manuscript and research into Māori astronomy, this book comes to life.

In recent times there has been a resurgence in Māori astronomy led by Māori working in this arena. Armed with precious manuscripts and oral histories, there is a new wave of Māori astronomical researcher working towards recreating the traditional whare kōkōrangi. Much of this new work is founded on te reo Māori, exploring traditional songs, incantations and any recorded account of Māori astronomy that makes up part of the language.[6] It also involves interviews with Māori cultural experts and holders of knowledge who have maintained star lore within their regions. This regeneration in Māori astronomy is part of a larger movement, where Māori themselves are striving to tell their own history and stories, in a manner that is acceptable and appropriate to Māori. Much of this knowledge is now being decolonised, and re-told by Māori based on Māori beliefs, Māori culture, Māori ways of thinking and Māori language.

To the forefront of this regrowth in Māori astronomy has been the celebration of Matariki. Since the early 1990s, the pre-dawn rising of Matariki in the month of June has been honoured as a time of togetherness, unity and goodwill. During this midwinter period, many groups coordinate events across the country to celebrate Matariki. These events have varied from the reciting of incantations on hilltops at dawn to balls, exhibitions, dinners and all manner of occasions. The rebirth of the Matariki celebrations has done much to highlight Māori astronomy, and to re-establish the Matariki tradition throughout the country. However, what is apparent is that there remains some confusion and misconceptions regarding the traditional observation of Matariki. While many wonderful events take place to commemorate Matariki, there are questions often posed in relation to its observation, the most frequent including:

- What is Matariki?
- Why did Māori observe Matariki?
- How did Māori traditionally celebrate Matariki?
- When and how should Matariki be celebrated?

This publication is concerned with answering the above questions, and with exploring the traditional customs associated with Matariki. By combining years of research with interviews of Māori cultural experts, this book attempts to explain what Matariki was in a traditional sense, in order for it to be fully understood and maintained within our contemporary society.

MATARIKI/PLEIADES

Matariki is more commonly known throughout the world as Pleiades or Messier 45 (M45).[1] This open star cluster is located in the constellation of Taurus the bull, near its shoulder. There are actually several hundred stars in Pleiades, of which only a handful are visible with the naked eye. Pleiades has an apparent magnitude[2] of 1.6, meaning it is not as bright as many of the other stars in the sky. However, it is immediately recognisable due to its uniqueness and beauty.

In Aotearoa, Pleiades is visible for most of the year, except for a month-long period when it sets in the west during the early evening of May, until its rise again in the pre-dawn sky during June or July. The easiest technique by which to find Pleiades is to look for other identifiable star groups and to use them as markers. One of the most familiar assemblies of stars is Orion's Belt, also known as the Pot, or Tautoru to Māori.[3] Nowhere else in the night sky will three stars of this brightness be seen so close together. If these stars are followed in a line from right to left, the observer will come to a triangular-shaped cluster of stars known as Hyades or the face of Taurus the bull. Māori call this group Te Kokotā[4] or Mata Kaheru.[5] Again, if the observer continues left, they will notice a small group of stars clustered closely together. This is Pleiades; this is Matariki.

Matariki can be located by using other well-known stars such as Tautoru. This image shows Matariki rising in the east before the sun during midwinter. Image: Te Haunui Tuna 2016.

PLEIADES IN THE ANCIENT WORLD

The Pleiades cluster is a prominent astronomical and cultural figure for many peoples throughout the world, and since early times its appearance has been observed, celebrated and worshipped. Palaeolithic drawings in the Lascaux Caves in southwestern France, estimated to be up to 20,000 years old, are said to represent the stars of Pleiades.[6]

Likewise, a group of seven stars on the 3600-year-old Nebra sky disk, discovered in Germany in 1999, is also thought to depict Pleiades. In China, the Pleiades are recorded as the 'Blossom Stars' in a 2357 BC reference that is regarded as the first piece of astronomical literature.[7] The Pleiades are recorded by Greek poet Hesiod around 700 BC and are documented by Homer in both *The Iliad* and *The Odyssey*.[8]

The cluster's most familiar name, Pleiades, is derived from ancient Greek and means 'sailing ones'.[9] This name reveals the cluster's close connection to the early mariners of the Mediterranean Sea.[10]

Another Greek version of the name is spelt 'Peleiades', meaning a 'flock of doves'.[11] In all, the Greeks had nine names for the different stars in Pleiades. Firstly, Atlas and Pleione, who are the parents of the other seven stars in this group. Their children are Electra, Maia, Taygete, Alcyone, Celæno, Sterope and Merope, all of whom are female, hence Pleiades is also known as the Seven Sisters.[12]

The Romans referred to Pleiades as 'Vergiliae' or 'Virgins of Spring',[13] and at times as a bunch of grapes. Elsewhere in Europe the Russians, Polish, Czechs and Slovaks called these stars the 'Hen and Chicks',[14] while the Vikings knew them as 'Freya's hens'.[15] Within Celtic traditions, the pre-dawn appearance of Pleiades after the long, cold winter marked the beginning of summer and the return of life to earth. This coincided with the Beltane Festival, or the May Festival, where bonfires were lit and people danced around the maypole.[16] The Celts also associate Pleiades with the honouring of the dead.

> The Celts also used the acronychal rising of the Pleiades to mark their month of mourning for dead friends. Prayers for the dead were said on the first day of what we now know as November. This custom is still echoed today with All Hallows Eve (October 31), All Saints Day (November 1), and All Souls Day (November 2), still celebrated as the feast days of the dead …[17]

In Mesopotamia, the Pleiades were known as 'Mul', meaning 'star',[18] and across the Arabic world they were referred to as 'Al Thuraya' or 'the many little ones'.[19] In ancient Egypt there were two main myths connected to Pleiades. To some they were called 'Chu' or 'Chow' and represented the goddess Nit or Neith. For others they were believed to be the stars of Hathor, the cow goddess, their symbol being seven cows.[20]

The Japanese name for Pleiades is 'Subaru', meaning 'gathering together'. It was a symbol of fertility, the changing seasons, and planting and harvesting, especially for rice.[21] The star cluster gets the name 'Flower Star' because it reaches conjunction with the sun in

Top: Paleolithic art of bulls and deer, on calcite walls of Lascaux Cave. Montignac, Dordogne, France.
Image source: Sisse Brimberg & Cotton Coulson, Keenpress/Getty Images

Bottom: The Nebra sky disk, discovered in Germany in 1999, is thought to depict stars including Pleiades.
Image source: Dbachmann via Wikimedia Commons, https://commons.wikimedia.org/wiki/File:Nebra Scheibe.jpg

The Greek names for the various stars within the Pleiades cluster.
Image sourced and adapted from http://hubblesite.org/newscenter/archive/releases/2004/20/image/a/.
Courtesy: NASA, ESA and URA/Caltech.

mid-May when the flowers are in bloom.[22] In Hindu tradition, Pleiades is 'Krittika', the seven daughters of Brahmā and Savitri. These seven daughters are said to have married seven wise men.[23]

There are numerous traditions associated with the Pleiades in the Americas. The Cherokee believe they descend from Pleiades,[24] and their solar year began with the heliacal setting of Pleiades around 1 November.[25] In Navajo legend, Pleiades is known as 'Dilyéhé', believed by some to be seven warrior boys. Dilyéhé was used by the Navajo to determine when to plant corn.[26] The Maya of Central America believe that Pleiades is part of a giant cosmic snake that connects the earth with the Milky Way.[27] For the Inca, the Pleiades played a significant role in their calendar system. Their mythology states that if this cluster appeared brightly in the sky, a bountiful harvest would follow; however, if seen small and indistinct, then a lean and hard year was at hand.[28]

Urania's Mirror showing Taurus the bull and the stars of Pleiades. Sidney Hall 1825. Image source: Sidney Hall [public domain] via Wikimedia Commons, https://en.wikipedia.org/wiki/File:Sidney_Hall_-_Urania's_Mirror_-_Taurus.jpg

One version from the aboriginal of Australia states that the Pleiades are seven women who are pursued by mythological men;[29] another account suggests the frosts of winter originate from the Pleiades, after seven sisters who were made from icicles fled into the sky. Once a year they pull icicles from their bodies and throw them to earth.[30]

Pleiades is mentioned on three occasions within the Bible, twice in the book of Job, Job 9:9 and Job 38:31, and again in the book of Amos, where the text reads:

> He who made the Pleiades and Orion
> And changes deep darkness into morning,
> Who also darkens day into night,
> Who calls for the waters of the sea
> And pours them out on the surface of the earth,
> The Lord is His name.

Throughout the ancient world the Pleiades star cluster was associated with changing seasons, death, planting, harvesting, weather, religion and new

Image of the seven sisters called 'Dance of the Pleiades' by American painter Elihu Vedder 1885.
Image source: Elihu Vedder [public domain] via Wikimedia Commons, https://en.wikipedia.org/wiki/Elihu_Vedder#/media/File:Elihu_Vedder_-_The_Pleiades,_1885.jpg

life, and was embedded into the beliefs and traditions of many indigenous peoples. This situation is also reflected in the cultures of the people who occupy the many islands of the Pacific Ocean, in particular Polynesia.

PLEIADES IN POLYNESIA

There is an undeniable connection that exists between the peoples of Polynesia, and this is echoed in the similarities in language, culture, sailing, art, religion and practice. The observation of Pleiades throughout the many island groups of this ocean expanse further reaffirms this cultural association. In the Hawai'ian archipelago, Pleiades is known as 'Makali'i',[31] and is connected to food and navigation.[32] One account states that Makali'i means the 'eyes of the chief' and the setting of Makali'i in the pre-dawn sky in November begins a four-month celebration known as 'Makahiki'. This festival honours Lono, the god of fertility, agriculture and peace, and includes organised games and events.[33]

An altar on the island of Kaho'olawe in Hawai'i showing sacred food that has been offered to the god Lono as part of the Makahiki celebrations. Photograph: Sam Kapoi 2016.

In Samoa, Pleiades is known as 'Li'i' or 'Mataali', also meaning 'eyes of chiefs';[34] in Tonga it is 'Matali'i', and the year is divided into two seasons based on this star cluster. When Pleiades appears on the horizon in the evening, the season is called 'Matali'i i nia' (Pleiades above), and when it is no longer visible after sunset, the season is called 'Matali'i i raro' (Pleiades below).[35] This same custom is followed in Tahiti, where Pleiades is known as 'Matari'i'.[36]

Rapa Nui (Easter Island) forms the most easterly point of the Polynesian triangle, and on this remote island, Matariki is also the name given to Pleiades. Its heliacal rising marked the new year, and when it disappeared from the sky in mid-April the fishing season came to an end.[37] Matariki is the name given to Pleiades within the Cook Islands; its rising was met there with feasting and celebration. Within the Mangaia section of this island group, Matariki had a deep connection to the ruling hierarchy. Sissons notes this association when he writes:

> It is also certain that the Mangaian chiefly hierarchy closely associated itself with Pleiades. Mangaia, the 'fish of Rongo', was (and remains) divided into six districts corresponding to the six visible stars of Pleiades. When a new high chief ('temporal Lord') was installed after a period of war, six processions were made around the island. Six layers of Rongo's sacred bark-cloth (*tikoru*) were wrapped around the high-priest of Rongo during his installation rite.[38]

In the Tuamotu archipelago it is known as Mata-ariki, translated as 'the eyes of the god' or 'the eyes of the chief';[39] on the coral atoll of Pukapuka in the Cook Islands it is called 'Mataliki';[40] the people of the Marquesas know this star cluster as either 'Matai'i' or 'Mata'iki'.[41]

Clearly Pleiades is an institution throughout Polynesia, and while the cultural practices surrounding its observation vary from island to island, the philosophical underpinnings and beliefs are much the same. Similar to the rest of Polynesia, Māori refer to Pleiades as Matariki; its influence over traditional Māori society was immense.

Northern Mariana Is
Marshall Is
Palau
Matariki
Kapingamarangi
Matariki
Kiribati
Matariki
Matariki
Nuguria
Takuu
Nukumanu
Solomon Is
Matali'i
Tuvalu
Mataliki
Papua New Guinea
Sikaiana
Matariki
Santa Cruz
Uvea
Rotuma
Wallis
Samoa
Anuta
Futuna
Tikopia
Vanuatu
Matariki
Fiji
Tonga
Matariki
Mataliki
Mataliki
New Caledonia
Kermadec
Matariki
New Zealand
Chatham

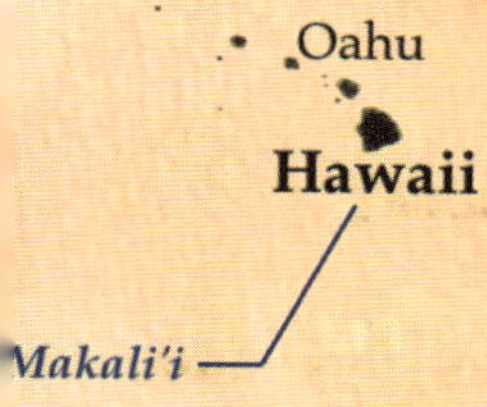

Map showing various islands in the Pacific and the names that these locations apply to Pleiades. Image: Te Haunui Tuna 2016.

THE MEANING OF MATARIKI

Matariki is often translated as 'little eyes' or 'small eyes', as Elsdon Best states: 'This looks as if Matariki was the name of a single star that has offspring six, but the plural riki in the name is against such a supposition (mata riki = small points or small eyes, the singular of small is iti).'[1] In Aotearoa, the translation of Matariki as 'little eyes' seems to have originated with Best, and subsequent authors have continued to support this position.[2] However, any greater meaning behind this name is left unexplained. This is because Best gives a literal translation of the word 'Matariki', and his interpretation lacks any greater connection to Māori understandings of cosmology.

There are scattered examples throughout Polynesia where Matariki is said to mean 'little eyes'; for example, in Mangareva it is sometimes called 'Matariki tinitini', or 'many little eyes'.[3] Still, the more common Māori translation is 'eyes of the god' or the 'eyes of the chief', and this position is supported by Makemson, who writes, 'In the Polynesian tongue, Matariki, the name for the Pleiades is contracted from Mata-ariki ...'.[4]

A particular Matariki story written by Kate Clark in *Māori Tales and Legends* states that Matariki was once a single star, larger and brighter than any other. Tāne became intensely jealous of Matariki, and with the help of Aldebaran and Sirius he attacked this star, eventually breaking it into six pieces. In Clark's version this is how Matariki gets its name 'little eyes'; however, it seems likely that this account originates from Mangaia in the Cook Islands and is not part of the Aotearoa understanding of Matariki.[5]

In his manuscript, Rāwiri Te Kōkau reveals that Matariki is in fact a truncated version of the name 'Ngā Mata o te Ariki Tāwhirimātea', meaning 'the eyes of the god Tāwhirimātea'.[6] This name finds its origins amidst the turmoil that existed not long after the separation of Ranginui and Papatūānuku.[7] Of the pantheon of gods who were present when Ranginui was forced from Papatūānuku, only Tāwhirimātea (Māori god of winds and weather) disagreed, and after Tānemāhuta had separated his parents, Tāwhirimātea sought retribution and began a series of attacks on the other gods. (Tāne is the Māori god of the forest. He completed many tasks, including separating the sky from the earth and suspending the stars in the heavens.) All the gods cowered before the wrath of Tāwhirimātea, except Tūmatauenga. Tūmatauenga is the god of war and humanity and is therefore the quintessential warrior. After an epic battle, Tūmatauenga emerged triumphant and vanquished his brother Tāwhirimātea, who, defeated and anguished, decided to flee skywards to spend his days with his father. But before he departed, Tāwhirimātea plucked out his eyes, crushed them in his hands and threw them into the sky, in a display of rage and contempt towards his siblings. It was also a symbol of aroha from son to father, revealing the deep-seated sorrow and affection Tāwhirimātea felt for Ranginui.[8] The eyes of Tāwhirimātea stuck to the chest of Ranginui and there they remain to this day. This is Matariki, or, as White declares, 'Mata Ariki',[9] a shortened version of the phrase, Ngā Mata o te Ariki Tāwhirimātea. Pou Temara adds to this discussion, stating, 'He atua kāpō a Tāwhirimātea'; 'Tāwhirimātea is a blind god'.[10]

Tāwhirimātea continues to send his descendants, the winds, to earth. It is for this reason Māori believe that the winds are so unpredictable, because Tāwhirimātea has no eyes and he uses the winds to feel his way around the world, seeking revenge for the separation of his parents by periodically causing storms that ravage his siblings.

In 1860 *The Maori Messenger* records a lament for Governor Browne, likening his death to the defeat of Tāwhirimātea:[11]

Tera Matariki te whetu o te tau,
E whakamoe mai ra,
He homai ona rongo,
Kia atu au,
Ka mate nei au.
I te matapouri,
I te mataporehu,
O roto i ahau
E whakatutuki ana
Ki ta te rangatira,
Ki tana whakaironga
Kuru rawa i aku iwi,
Kohi rawa i aku kiko,
Poka rawa i aku karu,
Tare ana i waho ra
Whanake te hikihiki,
Haere purangi te hua
Ki runga,
Maka ki tawhiti.
Ki nga whenua ahau.
E hoa, e Kawana Paraone

Yonder is Matariki,
Star of the season,
Taking his rest.
He now sends a summons
For me to depart.
Overwhelm'd is my spirit,
And dark is my heart,
As I approach the chief
And look upon his carvings
Bruis'd are my bones,
Consum'd is my flesh,
And my eyes, pluck'd out,
Are hanging from their sockets.
Utter now the incantation,
And lift high the offering
For I shall soon depart
To a far and distant land.
Friend, Governor Browne

THE NAMES OF MATARIKI

Matariki is known by a number of names, including 'Te Huihui o Matariki', meaning 'the assembly of Matariki' or 'the cluster of

Matariki'; 'Te Tautari-nui-o-Matariki', meaning 'Matariki fixed in the heavens'; and 'Tāriki', an abbreviation of Matariki.[12] 'Aokai' is another name that refers to the connection Matariki has with food; 'Hoko' or 'Hokokūmara' was applied to Matariki due to its influence over the growing of sweet potato.[13] In a lament for Te Koriwhai of Ngāti Korokoro from the Hokianga, the names Mataroa, Matarohaki and Matawaia are recorded alongside Matariki. Best suggests that these are possible additional names for this star cluster.[14]

Māori use the name Matariki to describe the entire cluster of Pleiades as well as a single star within the group. Hamiora Pio of Ngāti Awa gave seven names for the various stars in Matariki:

> Pio, of Ngati-Awa, gave the names of the six prominent stars of the group as Tupua-nuku, Tupua-rangi, Waiti, Waita, Waipuna-a-rangi, and Ururangi. He makes a curious remark that may possibly mean that Matariki is the name of a single star of the group, in which case we have the name of seven. He says: 'I will now tell you about another ancestor in the heavens, one Matariki, and her six children.' He then gives the six names as recorded above.[15]

In the above statement Pio gives Matariki and six names for the stars in the cluster, but their Western equivalents are not identified. Te Kōkau gives Matariki as the largest star in the cluster (Alcyone). This star is also referred to as the 'kai whakahaere' or the 'conductor' of the others. Te Kōkau pinpoints the additional six: Tupuānuku (Pleione), Tupuārangi (Atlas), Waitī (Maia), Waitā (Taygeta), Waipunarangi (Electra) and Ururangi (Merope). Furthermore, Te Kōkau identifies two extra stars within Matariki, taking the number in this group to nine. These stars are Pōhutukawa (Sterope) and Hiwa-i-te-rangi (Calæno). The star Hiwa is identified by Best as a descendant of Matariki.[16]

The notion of more than seven stars in the Matariki cluster is supported by the observations of the early New Zealand missionary

The nine stars of Matariki as recorded by Rāwiri Te Kōkau.
Image sourced and adapted from http://hubblesite.org/newscenter/archive/releases/2004/20/image/a/.
Courtesy: NASA, ESA and AURA/Caltech.

William Colenso, who noted the astonishing ability of Māori to see a greater number of stars with the naked eye than their Pākehā counterparts. He reported that Māori were able to see the satellites of Jupiter and 'not only seven stars of the Pleiades, but also several others'.[17]

Most cultures throughout the world speak of either six or seven visible stars within the Pleiades. However, this number can vary depending on certain factors, including the eyesight of the observer, location and light and air pollution. Different ancient cultures have records of six, seven, eight, nine, ten, eleven, fourteen and even sixteen visible stars within

Pleiades.[18] The Prophet Muhammad is said to have been able to see twelve stars in this cluster with his naked eye,[19] and as already noted, the Greeks saw nine and named them accordingly.

Matariki is identified in a number of sources as a female and the mother of the other stars in this group. In some accounts Matariki is referred to as a mother surrounded by her six daughters.[20] Other interpretations say Matariki are the Māori seven sisters and at times they are even described as a flock of birds. These versions are doubtful and seem suspiciously similar to the Greek myth where Pleiades are said to be sisters who were turned into doves and then into stars.[21] While Matariki is undoubtedly female, there is some conjecture surrounding the sex of her children. The following genealogy, as recorded by Te Kōkau, shows Matariki and her eight children, five of whom are females and three males. It also reveals that the father of the children is Rehua, great lord of the stars, the star Antares, which Māori believe is paramount chief of the heavens.[22]

The different names of the stars in the Matariki group are significant for Māori, as each individual has a defined purpose and is intrinsically connected with the Māori world. Their unique roles can be seen in the names applied to each.

The star Matariki was taken as a wife by Rehua and she gave birth to the eight children as listed above. Matariki is connected to well-being, and at times Matariki was viewed as an omen of good fortune and health. If the cluster, and especially the individual Matariki star, was seen high and bright in the night sky, it denoted good luck, peace and well-being for those who observed it. If it was seen in the sky when a patient was suffering from an illness, it was taken as a sign that they would soon recover. This belief is recorded in the following account:

> … the wizard doctor could revive a man if there were certain favourable conjunctions at the time; thus if the robin (toutouwai) was to sing for the first time just as the morning star (Tawera) was seen, if also the

Genealogy of Rehua, Matariki and their children.
Image: Te Haunui Tuna 2016.

> Pleiades (Matariki) were high in the sky, and the dying man had a shivering fit, then with all these auspicious signs occurring a certain invocation would bring back the departing soul.[23]

This association with health is reaffirmed in the saying 'Matariki, huarahi ki te oranga tangata'; 'Matariki, pathway to the well-being of man'.[24] It is important to note that Rehua is also connected to well-being and medicine, and those suffering from aliments would trust in Rehua for the power to heal:

> ... to the tenth heaven, where dwelt Rehua, the lord of Loving-kindness, attended by an innumerable host. Ancient of days was Rehua, with streaming hair. The lighting flashed from his arm-pits, great was his power, and to him the sick, the blind, and sorrowful might pray.[25]

It is within both Rehua and Matariki that knowledge of well-being and medicine exists, and both have the power to heal. Together, Rehua and Matariki produced the other stars in the cluster, each with its own unique purpose and meaning.

Pōhutukawa is connected to the dead, and in particular those who have passed from this world since the last heliacal rising of Matariki in the month of Pipiri.[26] Māori belief determines that when an individual dies, their spirit leaves their body and undertakes a journey along Te Ara Wairua, the pathway of the spirits. This journey ends at the northernmost point of the North Island at a place called Te Rerenga Wairua, the departing place of the spirits. The dead travel along the rocky ledge towards the ocean where an ancient pōhutukawa tree stands. They then descend down the aka (root) of this tree and disappear into the underworld. 'Below Te Aka, the long dry root of the pōhutukawa which does not quite reach the sea, is Maurianuku, the entrance to the underworld.'[27]

Pōhutukawa is the star that connects Matariki to the deceased, and it is the reason people would cry out the names of the dead and weep when Matariki was seen rising in the early morning.[28]

Pōhutukawa tree in bloom.
Photograph: Erica Sinclair 2016.

Tupuānuku is connected to food grown in the ground, and the word itself can in this circumstance be segmented, to get a clearer understanding of its intent. 'Tupu' or 'tipu' means 'to grow', and 'nuku' is a shortened version of Papatūānuku or earth.[29] Therefore, Tupuānuku means to grow in the earth. This star is connected to all cultivated and uncultivated food products and is the reason the Matariki cluster is immortalised in the proverb 'Hauhake tū, ka tō Matariki'; 'Lifting of the crops begins when Matariki sets'. When Matariki sets in the western sky at dusk during the month of May, the harvesting of the gardens has been completed and winter is near.[30]

Tupuārangi is similar to Tupuānuku in terms of its connection to food and growth; however, in this instance it is associated with food that comes from the sky. Tupuārangi is linked to birds. During the rising of Matariki, kererū were harvested in large numbers, cooked and then preserved in their own fat. This activity gave rise to the statement

Elderly Māori man sorting kūmara.
Northwood brothers: Photographs of Northland. Ref: 1/1-006227-G. Alexander Turnbull Library, Wellington, New Zealand.

Large kererū resting in tree.
Photograph: Erica Sinclair 2015.

'Ka kitea a Matariki, kua maoka te hinu,' meaning 'When Matariki is seen, the fat of the kererū is rendered so the birds can be preserved.'[31] Matariki, and in particular Tupuārangi, is the star that connects the cluster to the harvesting of birds and other elevated food products, such as fruit and berries from the trees.

Waitī means to be sweet,[32] and the saying 'he reo waitī' is applied when admiring a person with a melodious voice. 'Wai' is the Māori word for water, and in the case of stars, Waitī is connected to fresh water and all of the creatures that live within rivers, streams and lakes. The association Waitī and Matariki have with the animals of fresh water

Adult korokoro at Waikawa in the South Island.
Photograph: Dean Whaanga 2010.

is reflected in the proverb 'Ka kitea a Matariki ka rere te korokoro'. The korokoro is the lamprey, and these aquatic creatures leave the ocean during late winter and early spring, migrating up freshwater streams to spawn. This process occurs when Waitī is seen in the morning sky.[33]

Waitā is associated with the ocean, and in this situation the Māori word 'tā' means salt. It is similar to the word 'tōtā', meaning 'sweat'; hence Waitā is salt water. Waitā is the star that represents the many kinds of food Māori gather from the sea. Furthermore, it is said that when Matariki sits just above the water horizon, it has significant influence over tides of the ocean and the floodwaters. 'Te Waka Kawatini explained the Matariki (the Pleiades) has much influence in the control of tides when the star group is above the Tuahiwi nui o Hinemoana, the central ridge of the Ocean Maid, where the flood tides meet.'[34]

Māori woman fishing in Northland 1910.
Northwood brothers: Photographs of Northland. Ref: 1/1-006322-G
Alexander Turnbull Library, Wellington, New Zealand.

Waipunarangi is connected to rain, and the name itself can be translated to mean 'water that pools in the sky'. The pooling of water on the ground caused by the heavy and persistent showers of the winter months are referred to as 'Matariki tāpuapua'.[35] It is the star Waipunarangi that links the entire Matariki cluster to rain.

Top: The rain and pools of Matariki.
Photograph: Erica Sinclair 2015.

Bottom: The Tino Rangatiratanga flag blowing in the wind.
Photograph: Erica Sinclair 2016.

Ururangi means 'the winds of the sky'; a translation of the word 'uru' is 'west wind' or 'north-west wind'.[36] This star determines the nature of the winds for the year.

Hiwa-i-te-rangi is the final star in this group, and its name is connected to the promise of a prosperous season. The word 'hiwa' means 'vigorous of growth', and it is to Hiwa that Māori would send their dreams and desires for the year in the hope that they would be realised. This tradition is similar to the notion of wishing upon a star, or making a new year's resolution. In one particular source, Hiwa is said to have been a daughter of Matariki, and was taken as a wife by another star, Ioio-whenua. 'A wondrous genealogy shows that Ioio-whenua took to wife one Hiwa, a daughter of Matariki (the Pleiades), and that Tangaroa-i-te-rupe-tū, father of Maui, was a descendant of his.'[37]

Te Kōkau states that along with the names of the various stars in Matariki, there is also significant meaning in where they are positioned within the cluster. As the image on page 35 shows, Tupuārangi is above Tupuānuku, and this correlates with Māori understanding of cosmology, where Rangi the sky is always situated above Papa the earth. These two stars are associated with food, and there is both a male and female element, denoting balance.

Waitī and Waitā are linked to food and water; one is male and one is female. The reason Waitī is situated above Waitā is because fresh water always flows down to salt water, hence their positions in the sky.[38]

Waipunarangi and Ururangi are related to weather, being the stars that determine the rains and the winds. Again there is a gender balance, and both are situated above the other stars because this is where the rains and winds originate.

The purpose for the positions of Pōhutukawa and Hiwa-i-te-rangi is not recorded; however, Te Kōkau declares that these two female stars

are the most sacred of the group, as one is connected to the dead and the other deals with the deepest desires of the heart.

Matariki, the mother of these interstellar children, is placed in the centre, guiding her children across the sky, night after night.

Top: Matariki and her children in the sky.
Image: Te Haunui Tuna 2016.

Left: Te Rerengakōtukutahi Nikora looks to Hiwa-i-te-rangi in the early morning sky.
Photograph: Erica Sinclair 2016.

MATARIKI NEW YEAR

> Ana, i te atapō tonu ka rewa ake a Matariki ka kitea mai, ā koirā te tohu o te tau hou.[1]
>
> Therefore, in the early morning when Matariki is seen rising, this is the sign of the new year.

In order to understand Matariki and the celebration of the Māori New Year, it is important to have a basic understanding of maramataka Māori (the Māori lunar calendar), Māori seasons and the Māori year. For Māori, the year was divided into different seasons, month and nights. These were determined by a number of factors, including the position of the sun, the phase of the moon, the rising and setting of stars and ecological changes in the environment.[2] One major misconception is that Māori followed a universal lunar calendar and collectively observed the moon phases and seasons in the same manner. Māori understood that seasons, lunar phases and time itself is relative according to your location, environment and resource; therefore, the names and number of seasons, months and lunar phases differed depending upon region and tribe. Even the correct date to start the month could vary between tribal groups, and while most began with the new moon, others started their lunar month with the full moon.[3]

However, most Māori followed the lunar-stellar calendar that reckons the time and month of the year by observing the moon in relation to stars.[4] The position of stars in conjunction to the rising and setting of the sun was also observed by Māori, but it was the lunar calendar and not the solar calendar that guided the year. This belief is supported by Savage, who writes, 'The chief object of their adoration are the sun and the moon … the moon however is their favourite deity'.[5]

Generally, Māori observed four seasons and divided the year into twelve months. The seasons were kōanga (spring), raumati (summer), ngahuru (autumn) and hōtoke (winter). The following twelve names and explanations for the different months of the year were given to Elsdon Best by Tūtakangāhau of Tūhoe in the late 1800s:

1. Pipiri. *Kua piri nga mea katoa i te whenua i te matao, me te tangata.* All things on earth cohere owing to the cold; likewise man.
2. Hongonui. *Kua tino matao te tangata, me te tahutahu ahi, ka painaina.* Man is now extremely cold, and so kindles fires before which he basks.
3. Hereturi-koka. *Kua kitea te kainga a te ahi i nga turi o te tangata.* The scorching effect of fire on the knees of man is seen.
4. Mahuru. *Kua pumahana te whenua, me nga otaota, me nga rakau.* The earth has now acquired warmth, as also have herbage and trees.
5. Whiringa-nuku. *Kua tino mahana te whenua.* The earth has now become quite warm.
6. Whiringa-rangi. *Kua raumati, kua kaha te ra.* It has now become summer, and the sun has acquired strength.
7. Hakihea. *Kua noho nga manu kai roto i te kohanga.* Birds are now sitting in their nests.
8. Kohi-tatea. *Kua makuru te kai; ka kai te tangata i nga kai hou o te tau.* Fruits have now set, and man eats of the new food products of the season.
9. Hui-tanguru. *Kua tau te waewae o Ruhi kai te whenua.* The foot of Ruhi now rests upon the earth.
10. Poutu-te-rangi. *Kua hauhake te kai.* The crops are now taken up.

11. Paenga-whawha. *Kua putu nga tupu o nga kai i nga paenga o nga mara.*
    All haulm is now stacked at the borders of the plantations.
12. Haratua. *Kua uru nga kai kai te rua, kua mutu nga mahi a te tangata.*
    Crops have now been stored in the store pits. The tasks of man are finished.[6]

It is important to note that the different names for the Māori months are actually stars, and their heliacal rising in conjunction with the new moon denotes the beginning of the month. In 1854, Mahupuku stated that particular stars were placed in the Milky Way as markers of the various seasons.[7] Another version of the Māori year that is aligned to the morning rising of stars is as follows:

Te Tahi o Pipiri	The first of Pipiri	Pipiri or Pipirioterangi and the two stars Hamal and Sharatan in the constellation of Aries	(June)
Te Rua o Takurua	The second of Takurua	Takurua is Sirius in Canis Major	(July)
Te Toru Here Pipiri	The third of Toru Here Pipiri	Te Toru Here Pipiri is ζ Perseus	(August)
Te Whā o Mahuru	The fourth of Mahuru	Mahuru is Alpha Hydra	(September)
Te Rima o Kōpū	The fifth of Kōpū	Kōpū is Venus in the morning	(October)
Te Ono o Whitiānaunau	The sixth of Whitiānaunau	Whitiānaunau is ϒ Leo	(November)
Te Whitu o Hakihea	The seventh of Hakihea	Hakihea is θ Centauri	(December)
Te Waru o Rehua	The eighth of Rehua	Rehua is Antares in Scorpius	(January)
Te Iwa o Rūhiterangi	The ninth of Rūhiterangi	Rūhiterangi is Alniyat in Scorpius	(February)
Te Ngahuru o Poutūterangi	The tenth of Poutūterangi	Poutūterangi is Altair in Aquila	(March)
Te Ngahurumātahi o Paengawhāwhā	The eleventh of Paengawhāwhā	Paengawhāwhā is ε Pegasus	(April)
Te Ngahurumārua o Haki Haratua	The twelfth of Haki Haratua	Haki Haratua is η Pegasus	(May)

It must be stressed that the solar year and the Gregorian calendar[8] that we follow today is incompatible with the lunar calendar applied by the ancestors of the Māori. The Gregorian calendar is a solar calendar based on a 365-day year, while the lunar calendar follows a 29.5-day month, which equates to a 354-day year.[9] This means there is a shortfall of eleven days every year between the lunar calendar and the solar calendar. As the months in the Western calendar do not align with the months in the lunar calendar, using the Gregorian calendar system to determine the rising and setting of Matariki is fundamentally flawed.

Also, the number of moon phases within the maramataka ranged between twenty-eight and thirty-two nights, depending upon tribe and region. Each various phase of the lunar cycle had its own purpose and meaning and was therefore given its own name.[10] Māori were aware of the problems with synchronising a 354-day lunar calendar with a 365-day solar calendar, and to remedy this issue they applied two courses of action: 'One ... is that they did as their Eastern Polynesian ancestors did, and added some moon nights to the year to maintain seasonal synchronicity. Another method was to include an additional thirteenth month every few years.'[11]

For many tribal groups, the new year began with the heliacal rising of Matariki in the month of Pipiri, the lunar month that occurs around June, and ended when Matariki set with the sun in the evening of Haratua, the last month of the Māori year, which occurs around May.

Māori understood that the stars rose earlier every evening, and that the stars of the summer were different to the stars of the winter. Likewise, they noticed that the sun rose in different locations during the year. In midsummer it appeared south-east and set in the west, resulting in long and warm days. In the winter the sun emerged north-east and set north-west, and the weather was often inclement, with shorter days and long cool nights. The Māori proverb 'Te pō tūtanganui o Pipiri', meaning 'the long division of night during winter', refers to this cooler period.

The sun and his two wives, Hinetakurua and Hineraumati.
Image: Te Haunui Tuna 2016.

Māori interpreted the changing position of the rising and setting sun as a man moving between two wives. The sun is said to have married both Hinetakurua, the winter maiden, and Hineraumati, the summer maiden.[12] Hinetakurua is the star Sirius, seen on the eastern horizon in the month of Pipiri before the sun rises, denoting the winter season, hence the word 'takurua' means 'winter'. Hineraumati is said to inhabit the earth and is personified in the warm soil that supports the productivity of the gardens in summer.[13] Subsequently 'raumati' means 'summer'. In the winter, the sun rises in the morning with Hinetakurua. His stay in the sky is brief at this time, and he fails to warm the earth to any great extent. In this season the sun is far from earth, spending time with his winter wife and their children.

Then during the winter solstice, Matariki rises in the morning just north of the sun, signalling to this great being that he has stayed long enough with his winter wife, and it is now time to return east, back to Hineraumati and their children. The point where the sun turns to journey back to the east is known by Māori as 'te takanga o te rā', 'the turning place of the sun'. This phrase can be used to describe both the winter and summer solstice.[14]

After te takanga o te rā, the sun rises more easterly every morning, and the days become longer and warmer, bringing to earth the bounty of spring and summer. Matariki is therefore said to be 'te whetū o te tau' (the star of the year) because it signals the new year and informs the sun that it is time to return to the earth.[15]

Generally, the most important time for Māori to observe the stars was the morning, just before the rising of the sun. Many of the stars that sit along the northern and eastern horizon in the early dawn were indicators of the seasons.[16] These stars were markers that were connected to the time of year, ecology, animals and activities, and even foretold the bounty of the impending year. Of the many stars that were used by Māori as indicators, none were more significant and celebrated than Matariki. This point is recorded in the following statement:

> Te tahi o Pipiri ka puta o tatou matua, tupuna, ki waho i te ata po, i te wha o nga haora ki te titiro i te putanga o nga whetu. No te mea kei aua whetu te mohiotia ai te tau pai te tau kino, kei tenei whetu kei a Matariki, ka nunui nga whetu he tau pai mo nga ika.[17]

> In the first month of the year our ancestors would venture outside in the early morning, around four o'clock, to see the stars rise. Because these stars informed them whether it would be a bountiful or lean year, and the main star is Matariki, if it were large then it would be a good year for fishing.

The sun rising north-east in midwinter and south-east in midsummer.
Image: Te Haunui Tuna 2016.

While it is widely accepted that the Māori year commenced with the rising of Matariki, there are variations of opinion about when Matariki was observed and the length of the Matariki period. A number of writers have stated that Matariki New Year begins with the new moon after the cluster is seen in the sky before dawn. Others suggest that it was a tribal or regional specific celebration and could be observed after the full moon, in the new moon or in periods in between.[18]

It is unlikely that the extensive and detailed knowledge Māori had of astronomy and the environment would have been interpreted and applied in such a loose and unstructured manner. In addition, the accounts suggesting the new moon was the time to observe Matariki are also implausible due to Māori beliefs associated with this period of the lunar cycle. In traditional Māori society, day-to-day activities were dictated by the lunar calendar.[19] Māori believe that the various lunar nights affect the world and all its inhabitants;

Hinetakurua rising in the pre-dawn sky over the snow-capped Takitimu Ranges in the South Island. To the left of Hinetakurua are the three stars of Tautoru (Orion's Belt). Photograph: Erica Sinclair 2015.

the phases of the moon influence the behaviour of humans, the activities of animals and even the environment itself. The word generally used by Māori for the new moon is 'whiro', and according to Wiremu Tāwhai:

> The night is named after the god Whiro, another child of the family of Ranginui and Papatūānuku. Whiro is one of the younger brothers of Tāne and Tangaroa. Whiro has his own personality, his own aura, his own godly role pre-ordained and bestowed upon him by the celestial powers of creation. This is Whiro: god of dubious intentions, sometimes of sinister purpose, sometimes the obnoxious one. The elders advise caution and care, whatever one does at this time. If carelessness occurs, trouble will be attracted: this is the nature of Whiro.[20]

Whiro is the god of darkness, illness, disease and the origin of all ailments that afflict the world.[21] Because Matariki was used as a sign to determine the bounty of the year, it is extremely unlikely that its observation and reading would have taken place during such a threatening time. However, the period leading into the new moon is regarded as the most productive and fertile time of the month. This period begins with a three-day or four-day segment, depending on your local lunar calendar, known as 'ngā pō o Tangaroa', 'the nights of Tangaroa'. According to Tāwhai, the Tangaroa period was the time to observe the morning rising of Matariki. This statement aligns with Te Kōkau, who reflects that Māori would take note of the lunar phase and then wait for Matariki in the mornings of Tangaroa. Matariki might already be visible in the sky before the Tangaroa nights, but no celebrations or offerings were made until the correct lunar phase was seen.[22] The following text from Māori newspaper *Te Toa Takitini* points to Tangaroa as being the correct period to view Matariki:

> Ko Matariki kei Papa whakatangitangi e whitu nga po ki reira ka tae ki Maahu-tu, ka tae tenei ki nga po Tangaroa ko te tekau ma ono tenei o nga ra o Hune ka puta ake i te hiku o te Mangoroa.[23]

> Matariki is at Papa whakatangitangi for seven nights and then moves to Maahu-tu, on the nights of Tangaroa, on the sixteenth of June, it appears in the tail of the Milky Way.

The viewing of Matariki was spread across the three or four nights of Tangaroa to give people a chance to take their reading for the year. Winter is often a time of inclement weather, and conditions were not always favourable for viewing the stars. Therefore this window of opportunity was available in the most productive time of the month to view the rising of Matariki, and when Matariki was seen on the first clear night during the Tangaroa period, the New Year celebrations commenced. In 1875, Ngā Puhi prophet and chief Aperahama Taonui wrote about the link between both the setting and rising of Matariki and the nights of Tangaroa:

> Ka rumaki Matariki, i nga Tangaroa o Mei, 16 nga ra, ka tao ki Maukahau; po 7 ki reira. Ka haere, ki Tararauatea, po 7 ki reira, ka haere, ka tae ki te Papa whakatangitangi; po 7 ki reira, ka haere, ka tae ki Titore Maahutu, po 7 ki reira. Ka puta Matariki, i nga Tangaroa o Hune.
>
> But it is known that Matariki has a home when he enters the Tangaroa days of the middle of the month of May; he is in Mau-ka-hau, seven nights there. Thence he goes into Tararauatea and is seven nights there. Then he goes into Papa-whakatangitangi, and is seven nights there. Then he goes to Titore-mohutu, where he is also seven nights. Then he reappears in the middle of the month of June (Tangaroa).[24]

This particular article highlights the time between the setting of Matariki in the month of Haratua and its rising in the month of Pipiri. It states that during the Tangaroa days of Haratua, Matariki sets and travels to four different destinations, spending seven days at each. This is approximately a twenty-eight-day period when Matariki is not visible in the sky until its early morning rise in Pipiri, and it would usually extend from the end of the Tangaroa period in Haratua until the start of the Tangaroa period in Pipiri. This proposition is further supported by comments made by Tūtakangahau:

> Tūtakangahau of Tuhoe clearly explained the fact that in the Mataatua district the appearance of the Pleiades on the eastern horizon before sunrise was the sign awaited as a token of the new year. He made a peculiar statement that looks as though the year in that district commenced, or sometimes commenced, in the middle of the lunar month.[25]

It is important to note that Māori observed both the setting and rising of Matariki. The more well-known and celebrated festival occurred when Matariki rose in Pipiri welcoming in the new year;[26] however, its setting in the early evening of Haratua was also an important event. The tohunga kōkōrangi would wait for the moon to be in the Tangaroa phase and then observe the western skyline at the going down of the sun. As the sun slowly disappeared below the horizon, laments were said to the dead, as Māori believed Matariki carried those who had passed during the year into the great abyss of darkness.[27] Matariki sets next to the sun, but it is not visible due to the glare of daylight. Still, Māori recognised Matariki had disappeared with the sun because of the position of other stars such as Tautoru seen in the early evening of Haratua. The link between the setting of Matariki and the dead is portrayed in a lament composed by Nawemata for her husband, who died in the battle of Ruapekapeka:

> Tera te whetu e,
> Kapohia ana mai,
> Ka rumaki Matariki.
>
> There is the star
> Shining above
> Setting with Matariki.[28]

The setting of Matariki during the Tangaroa period in Haratua marked the end of the Māori year. This was also the beginning of a phase called 'mātahi kari pīwai', and this name is sometimes applied to Haratua, the twelfth month of the Māori year.

> The Pleiades appear to have been the most important group in Maori star lore, inasmuch as commencement of the Maori year was marked by the helical rising of that group, and it is also visible in the twelfth month of that year. In the twelfth month, sometimes known as the Matahi-kari-piwai, this group sets in the evening, thus indicating the last month (May June) of the year, while its appearance on the eastern horizon just before sunrise, marks the beginning of the year.[29]

Mātahi kari pīwai gets it name from the actions of the crop diggers, who during this period dug up the pīwai or tubers that had been overlooked and left in the gardens after the harvest. Mātahi kari pīwai lasts from

Matariki setting with the sun.
Photograph: Erica Sinclair 2016.

the early evening setting of Matariki in Haratua, to its pre-dawn rising in Pipiri, a period of about four weeks. This phase is important within the Māori calendar, as mātahi kari pīwai denotes the end of one year and mātahi o te tau marks the beginning of the next.

> As a native put it: "Matariki is the principal chief of the stars, because it shows the commencement of the year, and also the end. It acts in this way, in the Matahi-kari-piwai it sets, in Matahi o te tau [first month of the year] it returns."[30]

Te Kōkau asserts that the setting of Matariki was observed in the early evening during the Tangaroa lunar phase in the month of Haratua. Then its rising was acknowledged a month later in Pipiri just before the sun rose and when the moon was again in Tangaroa. He concludes that the Matariki celebration period lasted from the Tangaroa lunar phase right up until Mutuwhenua, or the night before the new moon. This is about a week-long period that spans the last quarter of the lunar period. However, all celebrations ceased just before the rising of the new moon to prevent Whiro from attending the festivities.

Following the western calendar is problematic for both the setting and rising of Matariki. There is a misconception that the first new moon during the month of June, according to the western calendar, is the beginning of the Matariki period.[31] However, this is far too early, and Matariki will not be seen in the morning sky at this time. The correct timing for Matariki is later in the month, when the moon is in the last quarter. Following the Māori lunar calendar, the setting of Matariki occurs between mid and late May, and periodically early June. Also, the rising of Matariki usually happens from mid to late June, and in some years early July, depending on the moon phases. The important point is that the western calendar and Māori maramataka are different, and the Matariki period should not be aligned with a non-Māori system of time.

Te Matahi o te Tau wharenui, Te Araroa.
Photograph: Erica Sinclair 2016.

Matariki rising as seen from the porch of Te Matahi o te Tau.
Photograph: Erica Sinclair 2016.

Whiro
Tirea
Hoata
Tamatea a Ngana
Tamatea Kai-ariki
Tamatea Tūhāhā
Ōhua
Atua Whakahaehae
Turu
Oike
Korekore Tuatahi
Korekore Rawea
Tangaroa Whakapau
Tangaroa Whāriki Koikoi
Ōtane

The Māori lunar calendar.[32] Matariki was viewed during the Tangaroa phases, and the Matariki celebration began on Tangaroa-ā-mua and ended at Mutuwhenua. Image: Te Haunui Tuna 2016.

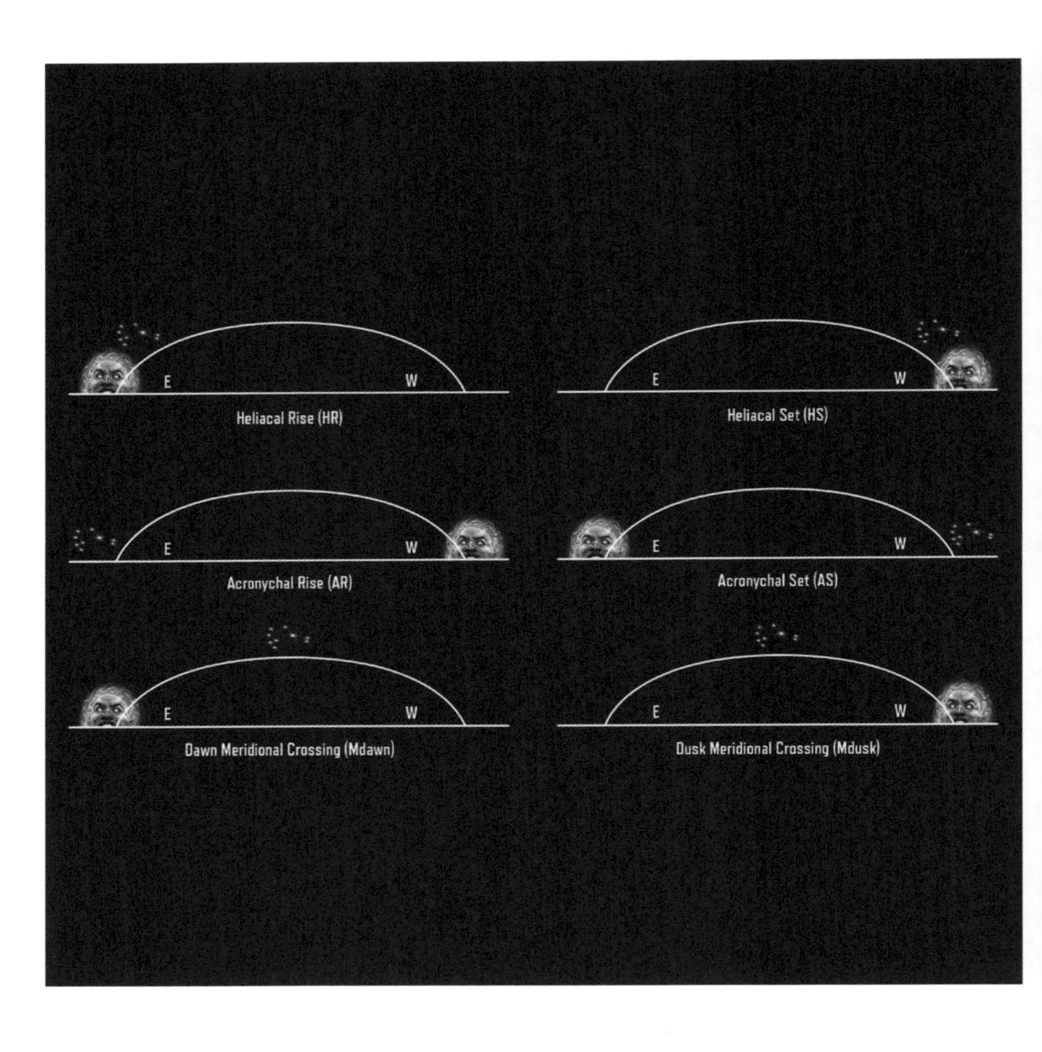

Stars were often observed in relation to the position of the sun. This image show how these observations worked.
Image: Te Haunui Tuna 2016.

RISING AND SETTING OF MATARIKI

Ancient cultures throughout the world often viewed stars in correlation with the lunar phase and the rising and setting of the sun, because the combined position of these celestial bodies enabled observers to determine the season.[1] Māori applied this technique when observing stars and, in particular, Matariki.

The image opposite shows the correlation between the position of stars and the rising and setting of the sun; these periods are called heliacal, acronychal and meridional crossing.[2] The heliacal rise is when a star is seen on the eastern horizon just before the rising of the sun. The heliacal set is when a star is observed on the western horizon just after the sun has set. Acronychal rise is when a star is visible on the eastern horizon as the sun sets in the west, and acronychal set occurs when the star is setting in the west as the sun rises. Dawn meridional crossing is when a star is at its highest point before the sun rises, and dusk meridional crossing is when the star is in the same position after the sun has set. Matariki was mostly observed during its heliacal rising and setting. However, there are also records of it being viewed during its meridional crossing, as the follow passage highlights:

> When the rata (Metrosideros robusta) was in full bloom, and Mata-riki (Pleiades) had passed the meridian of the sky, and autumn was near, and

> when the kumara-crop had been taken up and placed in the storehouses, the young people of Awhitu determined to pay the intended visit to Mount Eden.[3]

Critical to the observation of the heliacal rising and setting of Matariki is apparent magnitude. In astronomy, stars are given their own apparent magnitude value, a number that astronomers use to measure the brightness of an object in the night sky as seen from earth. The lower the value of the object on the apparent magnitude scale, the brighter it is. Historically, stars were divided into six magnitudes: the brightest became first magnitude stars, and those that are barely visible with the naked eye became sixth magnitude.[4] Since the invention of the telescope, many more magnitude levels have been added to the measuring system; however, the original magnitude 1 to 6 stars remain.[5] As already seen, Matariki has an apparent magnitude of 1.6; therefore, it is classed as a third magnitude star.

Object	Māori Name	Apparent Magnitude
Sun	Tamanuiterā	-26.74
Full Moon	Rākaunui	-12.7
Venus	Kōpū	-4.89
Jupiter	Pareārau	-2.94
Sirius	Hinetakurua	-1.46
Canopus	Atutahi	-0.74
Rigel	Puanga	0.13
Antares	Rehua	0.96
Fomalhaut	Ōtamarākau	1.16
Pleiades	Matariki	1.6
Mintaka	Matatoki	2.23

When searching for stars on the eastern horizon before the sun rises, a number of factors become important, including the glow of the sun, the

Matariki at 5 degrees above the horizon with the sun 16 degrees below the horizon, meaning Matariki is visible in the sky.
Image: Te Haunui Tuna 2016.

position of the moon, atmospheric conditions, local topography and the apparent magnitude of the star.

> 1st magnitude stars become visible at heliacal rise/set at a minimum altitude of 5°... when the sun has an altitude of -10°. For 2nd magnitude stars, the sun's altitude needs to be a minimum of -14°, and -16° for 3rd magnitude stars.[6]

Since Matariki is a third magnitude star, for its heliacal rise to be visible it will need to be at least 5 degrees above the horizon while the sun is at least 16 degrees below.

Using the Māori lunar calendar, the position of the sun and apparent star magnitude, the following table shows the setting, rising and celebration period for Matariki until 2050.

Year	Setting of Matariki	Rising of Matariki	Matariki Period
2017	19 May	17–20 July	17–22 July
2018	7 June	6–9 July	6–13 July
2019	27 May	25–28 June	25 June–3 July
2020	15 May	13–16 July	13–20 July
2021	2 June	2–5 July	2–10 July
2022	23 May	21–24 June	21–29 June
2023	13 May	10–13 July	10–17 July
2024	31 May	29 June–2 July	29 June–6 July
2025	21 May	19–22 June	19–25 June
2026	8 June	8–11 July	8–14 July
2027	29 May	27–30 June	27 June–4 July
2028	16 May	15–18 July	15–21 July
2029	4 June	4–7 July	4–12 July
2030	24 May	23–26 June	23 June–1 July
2031	14 May	11–14 July	11–19 July
2032	1 June	30 June–2 July	30 June–8 July
2033	22 May	20–23 June	20–27 June
2034	11 May	9–12 July	9–15 July
2035	30 May	29 June–1 July	29 June–5 July
2036	18 May	17–20 July	17–22 July
2037	6 June	6–9 July	6–13 July
2038	26 May	25–28 June	25 June–3 July
2039	15 May	13–16 July	13–20 July
2040	2 June	1–4 July	1–9 July
2041	23 May	21–24 July	21–28 July
2042	13 May	10–14 July	10–17 July
2043	1 June	30 June–3 July	30 June–7 July
2044	20 May	19–22 June	19–25 June
2045	8 June	7–10 July	7–14 July
2046	28 May	26–29 June	26 June–4 July
2047	17 May	15–18 July	15–22 July
2048	4 June	3–6 July	3–11 July
2049	24 May	22–25 June	22–30 June
2050	14 May	11–14 July	11–18 July

Therefore, during the early mornings of Pipiri, when the moon was in the Tangaroa phase, Māori would wait for Matariki to appear in the eastern sky. It was believed that Matariki marked the changing of the year, as the following text suggests: 'Ko Matariki te whetu kei te arahi i tenei marama, he wehenga tau ki ta te Maori whakahaere.'[7] 'Matariki is the star that leads this month, and divides the year according to Māori understandings.'

The name of this first month of the Māori year was sometimes known as 'Mātahi o te tau', meaning the 'first month of the year', and because Matariki was the sign of this month, it was also known as te whetū o te tau. This title for Matariki is recorded in the follow lament:

Tirohia Matariki,
Te whetū o te tau,
E whakamoe mai rā,
E homai ana rongo

Look at Pleiades,
Star of the year,
Preparing to sleep up there,
It signals its news[8]

THE BOUNTY OF THE YEAR

The first appearance of Matariki in Pipiri during the Tangaroa period was an important time, as this cluster brought forward the bounty of the impending year.[9] Once it was sighted on the horizon, the astronomical experts of the village would keenly observe Matariki in great detail. Each of the nine individual stars would be assessed, and mental notes would be made about their brightness, distinctiveness, colour and distance from the surrounding stars. Likewise, the movement, colour and shape of the entire cluster would be noted. From these observations the tohunga kōkōrangi would make predictions about the productivity of the new year. As Best notes, '... if the stars of this group are indistinctly seen at the time of its heliacal rising, and they seem to quiver or move, then a cold season follows. If they are plainly seen at that time – stand out distinctly – a warm, plentiful season ensues.'[10]

This statement is endorsed by Riley, who writes, 'When these stars appear far apart, a warm and plentiful season is in prospect; if they appear close together, a cold and meager one'.[11] While these explanations are correct, and the Matariki cluster was viewed in its entirety, its signs and related readings were more detailed and precise than these quotes imply. The general nature of the following seasons could be predicted by viewing the cluster as a whole, and if it was seen to be hazy and shimmering in the morning sky, a poor season would ensue. However, if the stars were bright and clear and seemed to stand motionless in the sky, then a bountiful season would follow. This was also a sign that the weather would be warm and consistent.

Aside from this generic reading of Matariki, the different stars in this group would be observed, and a prediction would be applied to each. As already discussed, all of the nine stars in Matariki have their own names and each has an assigned purpose. If Tupuānuku appeared dim and small, the produce from the village gardens would not be as large or as plentiful as in previous years. Yet if this star was seen bright in the sky, then the storehouses would be full after the next harvest. Likewise, if Tupuārangi showed poorly in the sky, then the fowlers would struggle to secure large number of birds. Similarly, the appearance of Waitī and Waitā would inform the tohunga kōkōrangi about the bounty they could expect from the rivers, lakes and oceans in the following months. Waipunarangi is concerned with the rain, and if this star was indistinctive in the sky, then the yearly rainfall would be higher than usual, and flooding could be expected. If, however, it revealed itself in a bright and clear manner, then there would be less rain than usual. This is also the case for Ururangi, which is the star connected to the winds.

If Ururangi appeared dull when Matariki rose, a windy year would follow. If it were bright and distinct, then the wind throughout the year would be of a more pleasant nature.

Environmental readings were not taken from the stars Pōhutukawa and Hiwa-i-te-rangi. These two stars were significant for different reasons. As already seen, Pōhutukawa is linked to the deceased. If this star was easy to see in the sky, there would be few deaths within the community that year. If Pōhutukawa was seen to move erratically, or was invisible when the cluster rose, then the village could expect to mourn the passing of many people in the perusing months.[12]

It was to Hiwa-i-te-rangi that the hopes for the new year were sent, in the wish that they would become lush in the heavens and bring forth fruits for people on the earth. This is very similar to the idea of making a wish upon a star, in the hope that it would come true. Māori believed that if Hiwa-i-te-rangi were seen bright in the sky at New Year, then the individual and the collectives desires for the year would come true. This belief in Hiwa is captured in the following lines of a karakia:

Ko Hiwa
Ko Hiwa nui
Ko Hiwa roa
Ko Hiwa pūkenga
Ko Hiwa wānanga.
Takataka te kāhui o te rangi
Koia a pou tō putanga ki te whai ao
Ki te ao mārama.

Hiwa
Great Hiwa
Large Hiwa
Skilled Hiwa
Knowledgeable Hiwa.

Let the stars fall from the sky [my wishes]
And be realised in this world
The world of light.[13]

The appearance of the stars in Matariki can vary significantly from year to year, and this impacts on its reading. For example, Tipuānuku and Tupuārangi might appear clear and strong, however Waitī and Waitā might appear less impressive. This would be a sign that fishing and eeling would be less productive, and the community would increase its focus on growing vegetables and hunting birds. Ururangi might be bright but Waipunarangi less so, signalling that the wind would be less of a concern than the rain. A number of combinations could appear with the rising of Matariki, informing the people of the year ahead. It fell to the tohunga kōkōrangi of the area to determine the meanings held within the Matariki cluster, and then to share their predictions with the masses. The ability to make these judgements was not something taken lightly, for it would have significant implications for the entire year. It was also a skill that was not acquired in one viewing of Matariki, but was built up over many years in te whare kōkōrangi. Tohunga kōkōrangi would spend the evenings within the confines of te whare kōkōrangi, debating the universe and staring into the cosmos.[14]

MATARIKI, A SIGN OF DEATH

Ko Matariki te kaitō i te hunga pakeke ki te pō

Matariki draws the frail into the endless night[15]

Matariki is associated with death, hence the saying 'Matariki whanaunga kore, Matariki tohu mate', meaning 'Matariki the kinless, Matariki a sign of death'.[16] In particular, the star Pōhutukawa within the Matariki cluster is associated with those who have passed into the endless night. Once the tohunga kōkōrangi had ascertained the Matariki signs for the year, the community would gather together to mourn the dead.

Matariki would then be greeted with song, tears and lamentations – this particular ceremony is known as 'te taki mōteatea' or the 'reciting of laments'.[17] It included incantations to the dead and the calling of the names of those who had passed since the last rising of Matariki.

Matariki is said to be the prow of a great canoe in the sky. The generic name for this canoe is 'te waka o Rangi', or 'the canoe of Rangi'.[18] At times this canoe is regionalised, depending on where it is seen. For example, in the Waikato area it is known as 'te waka o Tainui', which is the ancestral vessel of this area; the canoe extends from Matariki to Tautoru and includes Te Kakau (Orion's scabbard). The captain of this canoe is Taramainuku, who is the owner of a giant net. When Matariki rises in the morning of Pipiri this canoe is seen sitting on the horizon, and Taramainuku stands on board with his net. For the next eleven months of the Māori year, from Pipiri until Haratua, Taramainuku casts his net across the earth and hauls to the sky all those who have died that day. The spirits of the deceased are suspended by Taramainuku to the stern of the canoe, at a place known as Te Hao o Rua (Orion's nebule). There they hang like the kura, the plumes of decorative feathers that adorn Māori canoes.

The image of Taramainuku pulling on the ropes of his net to ensnare the dead at the sighting of Matariki is seen in the following lament composed by Kauhoe of Taranaki for Te Whao and Tupoki:

Ka ripa ki waho ra, e Atu-tahi koā
Te whetū tārake o te rangi,
Ka kopi te kukume,
Ka hahae Mata-riki e,
Puanga, Tau-toru,
Nāna i kukume koutou ki te mate, ē

Away out yonder is Atu-tahi
The star that shines apart in the heavens,
The noose was pulled taut,

At the rising of Mata-riki,
In the company of Puanga and Tau-toru,
It was thus you were all hauled down in death, alas[19]

Throughout the year Taramainuku hauls the dead from the earth and places them on his canoe. Then, in the month of Haratua, te waka o Rangi is seen vertical on the western skyline, setting with the sun. Māori waited for the Tangaroa nights of Haratua when the year came to an end. Then they would farewell their dead as Matariki escorted them into the afterlife. The setting of Matariki and te waka o Rangi would be observed with tears, laments and sorrow. Even though Matariki was not visible setting so close to the sun, Māori could estimate its position due to the location of the stern of the canoe after the sun has set. The notion of Matariki leading the dead into the great abyss is encapsulated in the saying, 'te ope o te rua Matariki', 'the company of the cavern of Matariki'.[20] This proverb is in reference to the dead of the year, disappearing next to Matariki into the great cavern of darkness. This belief is also recorded in a section lament composed by Te Wheoro of Waikato at the passing of his daughter:

Te mate o te tangata.
Ka rū te whenua, ka rere Tautoru
Te rua o Matariki, ko te tohu o te mate.

Death is the lot of humankind.
The earth trembles, Orion's belt journeys on
The pit of Pleiades, the sign of death.[21]

When Matariki and te waka o Rangi return to the morning sky in Pipiri, they once more carry the dead of the previous year; however, they have now been prepared for their final journey to become stars in the sky. Taramainuku gathers the spirits of the year from the stern of the canoe and casts them into the heavens to become stars.

Many Māori believe that this is where the dead spend eternity, as stars adorning the night sky. This is also the origin of the well-known Māori phrase when mourning the deceased, 'kua whetūrangihia koe', 'you have become a star'. It was also this constellation that marked an auspicious occasion in 1906, as Kirkwood explains:

> The arrival of Te Ra o Tainui, Te Matariki and Pipiri also marked the period when the dead were remembered and on this night our tupuna would call out the names of those who had departed in the previous year. It was on this night in 1906, Koroki Te Rata Mahuta Tawhiao Potatau Te Wherowhero was born. He was the child chosen and destined to be the fifth Maori King. It is the most sacred of nights. It is the night to give thanks to the goodness of the past and karakia for the future well-being and the night that our tupuna remembered those who passed away.[22]

When Māori saw Matariki return as part of te waka o Rangi in the morning of Pipiri, they knew Taramainuku was casting the spirits of their kin into the heavens. Hence, they would mourn their loved ones and call their names. Tohunga would conduct karakia as part of this mourning process, reciting the names of the dead and sending their spirits on their way. Aspects of this traditional practice, of waiting for the return of Matariki and the year to turn before releasing the dead, are still evident in modern Māori culture. Often Māori will wait a year after the death of a close relative before they unveil a headstone or begin practising particular cultural obligations.[23] This present-day ritual finds its origins within the wider celebration of Matariki.

The intimate connection between Matariki and the dead has been immortalised within a number of Māori proverbs, songs and phrases. For instance, 'ka takitahi ngā whetū o te kāhui o Matariki' means 'the stars of Matariki are becoming fewer'. This metaphor compares the death of a person to the disappearance of one of the stars of Matariki. The person has passed into the night; therefore, one of the stars of

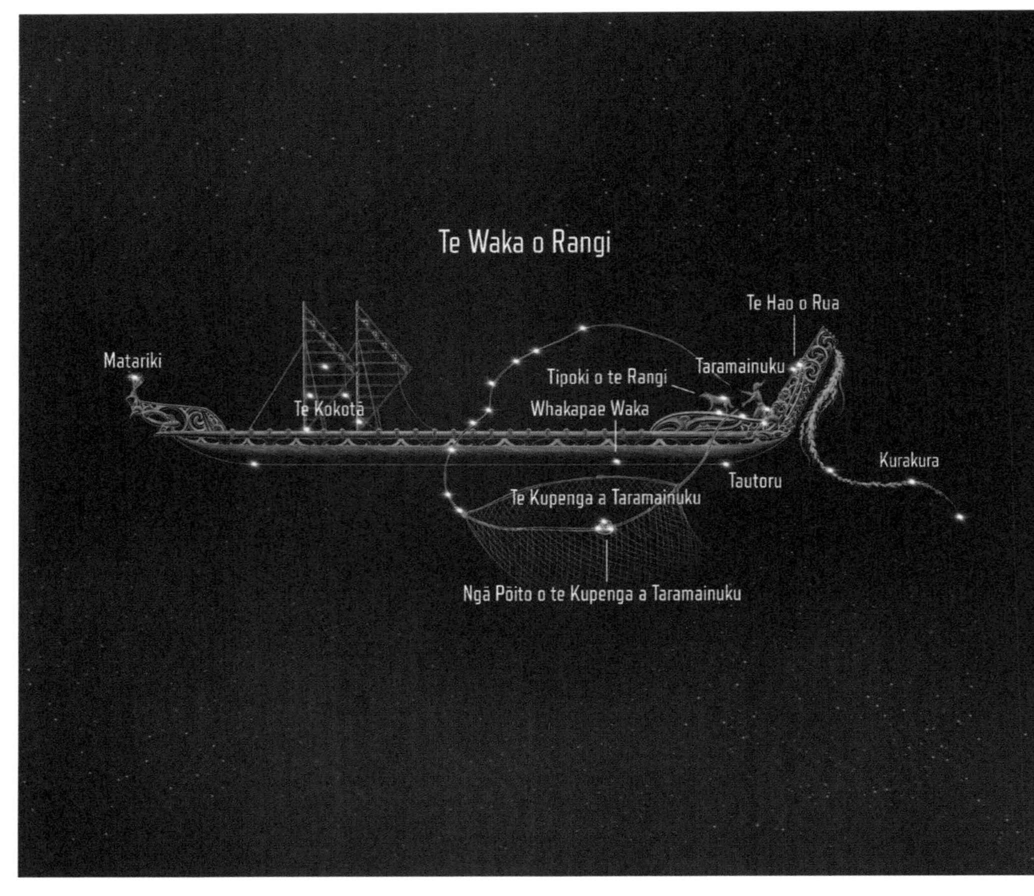

Taramainuku aboard his canoe, Te Waka o Rangi, casting his net below to haul skyward those who have passed during the day.
Image: Te Haunui Tuna 2016.

Matariki has been extinguished. This phrase was used during a eulogy to Tuini Ngawai, famous composer of Ngāti Porou.[24] Another example of how Matariki was used to express emotion, especially sorrow, is captured in a segment of a lament composed by Tūhoe composer Mihi-ki-te-kapua:

> Tirohia atu nei nga whetu,
> Me ko Matariki e arau ana;
> He hōmai tau i nga mahara
> E kohi nei, whakarerea atu
> Nā te roimata ka hua riringi
> Taheke ware kai aku kamo.

I gaze up at the stars,
And the Pleiades are gathered together;
Which gives rise to many thoughts
That well up within, and freely
Do the tears pour forth
And flow shamelessly from mine eyes.[25]

MATARIKI: WHĀNGAI I TE HAUTAPU

One of the important ceremonies that occurred at the first sighting of Matariki, after the bounty of the season had been ascertained and the mourning of the dead completed, was the practice of whāngai hau; 'an offering of ceremonial food to an atua'.[1] The term used to describe this ceremony, 'whāngai i te hautapu', means 'to feed with a sacred offering', and the word 'hautapu' means 'a sacred portion of food that is offered to the gods'.[2] Escorting the dead across the night sky, bringing the bounty of the year and signalling the return of the sun to earth takes a great toll on Matariki. In the early morning of Pipiri, Māori would offer unto this star cluster a selection of the best of the harvest, to reinvigorate Matariki for another year of service. This practice seems to have been common, and as Densey states:

> ... the late Mr Rangihuna Pire, of South Taranaki, who told me in 1957 – he was then in his 70's – that he used to be taken by his grandparents to watch for Matariki at night in mid-winter. That was at Kaupokonui, in South Taranaki. The old people might wait up several nights before the stars rose. They would make a small hāngi. When they saw the stars, they would weep and tell Matariki the names of those who had gone since the stars set, then the oven would be uncovered so the scent of the food would rise and strengthen the stars, for they were weak and cold.[3]

Leading into the Matariki period, the community would select a portion of different food items as an offering, which would represent the various stars in Matariki. For example, a kūmara would be acquired for Tupuānuku, and a kererū or another bird from the forest would be offered to Tupuārangi. Likewise, an eel or freshwater fish would represent Waitī, and some form of shellfish or sea life would be chosen for Waitā. Only the best and most appropriate foods would be offered, and these were carefully prepared. In the evening before the rising of Matariki, these items would be taken to the tohunga. Then at a ceremonial place, a special oven would be prepared, te umu kohukohu whetū, the 'steaming earth ovens of the stars'. After rocks were heated on a fire, they would be transferred to a small pit, with the uncooked food placed on top and covered with leaves and earth. At the completion of the incantations to Matariki and the calling of the names of the dead, the ovens were uncovered, and the steam from within would rise into the sky and replenish Matariki. Matariki would then gather the offerings from below and feast upon them. This action is celebrated in the proverb 'ngā kai a Matariki nāna i ao ake ki runga'. This phrase is often translated to mean 'the food of Matariki that is scooped up', because Matariki is a star that brings food. The word 'ao' means 'to scoop up with both hands', and in this situation the saying actually means, 'Matariki has accepted the offerings, and this will ensure it will return next year to bestow its bounty upon the world.'[4]

The steam that rises from the cooked food of te umu kohukohu is known as hautapu,[5] and this is the final segment of the early morning Matariki ceremony that opens the new year celebrations. This ritual ends with the rising of the sun. The use of hautapu to replenish the vitality of Matariki, and the releasing of the dead as stars before the rising of the sun, is documented in a section of a famous haka from the east coast of the North Island. This part of the haka is known as a 'tuku', which means 'to release, let go or relinquish', and it was traditionally part of a series of karakia that were conducted during the time of

The whāngai i te hautapu ceremony being conducted during the first Tangaroa night in Pipiri in 2016. The photograph was taken from Hakarimata overlooking Hamilton as Matariki appeared between the clouds. Photograph: Erica Sinclair 2016.

Matariki. C Company of the Māori Battalion performed this haka on 23 September 1941 at El Diyura in Egypt:

Ko te hautapu e rite ki te kai nā Matariki
Tapareireia koiatapa
Tapa konunua koiana tukua
Hi aue hi
I āhahā
Ihi ka tū te ihi, ihi ka tū te wanawana
Ki runga i te rangi e tū iho nei, tū iho nei
Hī

The sacred food that is prepared to replenish Matariki
Severing our bond to those who have departed
Releasing them
Alas

> Then rises the sun
> Higher and higher into the sky
> Completing the ceremony[6]

MATARIKI CELEBRATIONS

Once the ceremonies connected with the first sighting of Matariki had ended, a period of celebration ensued. Matariki was a time of rejoicing and celebrating: people came together to enjoy the company of friends and family. Māori believed that when Matariki gathers in the sky, it calls people to gather on earth. This belief is celebrated in the proverb 'Matariki hunga nui', meaning 'Matariki has many people', because Matariki gathers people together.[7]

Best states that a festival commenced at the rising of Matariki and this began a period of song, dance and feasting. This statement is supported by White, as cited in Hakaraia:

> The tapu period of the year was when Matariki appeared above the horizon in the morning. That was the occasion on which our elders of former times held festivals when the people rejoiced and women danced and sang for joy as they looked on Matariki.[8]

The rising of Matariki also coincided with a period of inactivity. The harvest had been completed, and the storage houses and pits were full with produce. Apart from the harvesting of kererū and korokoro, little work was undertaken at this time, and people were free to relax and entertain themselves with activities of a more pleasurable nature, such as dancing, music, art, games and other pastimes. These activities fall under the domain of four ancestors of the Māori, namely Raukatauri, Raukatamea, Marereotonga and Takatakapūtea. Interestingly, these four beings are recorded as circumpolar stars in the southern sky. They are associated with peace and entertainment, and were key figures during Matariki festival, as the following claim suggests: 'Raukatauri,

A group of children holding up a variety of Māori string patterns.
Making New Zealand: Negatives and prints from the Making New Zealand Centennial Collection
Ref: MNZ-2424-1/2-F Alexander Turnbull Library, Wellington, New Zealand.

Raukatamea, Marereotonga, and Takatakaputea. The arts of pleasure were in evidence more particularly when the crops had been lifted and stored, when a harvest feast and period of merry making ensured. This was the Pleiades festival.'[9]

The Matariki festival period began at the first sighting of the cluster during the Tangaroa phase of Pipiri and ended at the new moon. This is the last quarter of the lunar phase and lasts for approximately one week. This is the most productive and successful phase of the lunar calendar, and it was in this prosperous period that the Māori year was observed. Matariki celebrations ceased with the new moon, which for Māori is a low period, and a time to be cautious and aware of the dangers of the world.

This celebration phase is known as 'te mātahi o te tau', meaning 'the first fruits of the year', or 'the new growth of the year'.[10] This term is sometimes used to denote the first month of the year, but more specifically it is connected with the bounty of the year that is promised during the rising of Matariki. This was a time for celebrating the promise of new life and prosperity.[11]

MATARIKI AND PUANGA

In recent times, much has been made of the belief that the star Puanga (Rigel) is the marker of Māori New Year, and in various locations this is correct: Best noted that 'In some areas, such as the far north and the Chatham Islands, the rising of Puanga seems to have marked the turning of the year'.[12] This situation extends to parts of the west coast of the North Island and much of the South Island. This is not to say that these tribal groups don't have an association with Matariki, or that tribes who adhere to Matariki don't observe Puanga. Rather, it shows that the observation of the new year was relative to location and there was more than one marker of time. Often Māori observed a combination of indicators when distinguishing time, and stars like Pipiri, Puanga, Tautoru, Whakaahu and Matariki would be studied as a collective.[13] This belief is recorded in the following section taken from the Māori newspaper *Pipiwharauroa*:

> Te tahi o Pipiri ka puta o tatou matua, tupuna, ki waho i te ata po, i te wha o nga haora ki te titiro i te putanga o nga whetu. No te mea kei aua whetu te mohiotia ai te tau pai te tau kino, kei tenei whetu kei a Matariki, ka nunui nga whetu he tau pai mo nga ika. Kei aua whetu kei a Puanga kei a Matariki.[14]
>
> In the month of June our elders would go outside in the early morning at 4 a.m. to look at the stars. Because these stars would inform them if it was going to be a fruitful year or a lean year, if the stars were large, it would be a good year for fishing. The signs were with the stars Puanga and Matariki.

There is even a suggestion that for some regions Matariki is the sign that the old year has passed, and Puanga is the sign that the new year is upon us:

> Mo te marama o Hune ara o Te Tahi o Pipiri o te Tau, 1922. Ko Matariki te whetu o te tau tawhito kei te arahi mai i a Puangarua te whetu o te Tau Hou.[15]
>
> In the month of June, the first month of the year in 1922. Matariki is the star of the old year leading Puanga the star of the new year.

One of the misunderstandings around Matariki and Puanga stems from the belief that Puanga is celebrated in some regions because the people in these areas cannot see Matariki in the sky during Pipiri. As we have already seen, the Pipiri month and the 'correct' dates to view Matariki are often confused because of the difficulties of trying to sync the traditional Māori lunar calendar to the Western calendar. Modern interpretations of Matariki have seen this celebration occur too early, and may have given rise to the notion that in some areas Matariki cannot be seen, therefore Puanga is celebrated. However, the variation in rising between Matariki and Puanga is very small, and if the Tangaroa nights of Pipiri are observed correctly, then both stars will be seen in the morning sky.

Matariki and Puanga are often recorded together in songs and phrases; for example, 'Tākina mai rā ngā huihui o Matariki e; Puanga Tautoru māna e whakarewa te ika wheturiki, ka rewa kai runga.'[16] 'Rise together the stars of Matariki, of Puanga and Tautoru, setting in motion the smaller stars that rise above.'

Māori also believe that the relationship between these two stars is strained, and Puanga is said to be ever hostile towards Matariki.[17] This hostility originates from the desire of Puanga to be the star that heralds the new year. Therefore, Puanga spends time combing her hair and making herself beautiful; hence the name 'Puanga kakaho', meaning

'Puanga the fair-haired'. This is done to entice the sun to rise next to her in the month of Pipiri, marking her as the star of the year. Puanga repeats this process year after year, seeking the approval of the sun; hence the name 'Puanga raia', meaning 'Puanga the persistent'. However, in the month of Pipiri, the sun rises next to Matariki, signalling to the world that the year has turned.

The important point is the fact that all tribes maintained knowledge of both Matariki and Puanga, and it was the prerogative of the individual tribe, or community, as to what symbols and ecological events they chose to follow in order to determine the new year. This situation remains to this day, and while some groups see Matariki as the symbol of the new year, others choose instead to acknowledge Puanga.[18]

THE FOOD OF MATARIKI

Matariki has a deep association with food, from planting to growing, hunting to gathering, and food storage.[19] As already discussed, the various stars in the Matariki cluster and their appearance in the morning sky during Pipiri predicted the bounty of the following year. Hence, Matariki was referred to as a 'whetū heri kai'; a star that bestowed upon the earth the spiritual essence of food every year.[20]

Historically, many Māori communities would plant a sacred garden at the rising of Matariki. This garden was known as 'te māra tautāne'.[21] Te māra tautāne was separate from the general communal plantations and was the property of the gods; therefore it was sacred, and was planted with ceremonial seeds when Matariki rose in the winter, well before the main gardens were sowed.[22] The planting ceremony connected to these sacred crops is called 'huamata'. All the produce grown within te māra tautāne were offered to Rongo, the Māori god of cultivations, and to the star Matariki.

> I te Ao kōhatu i mua i te whakatōtanga o te iwi i ngā kai, ka rāhuingia Te Māratapu mā te Atua o te kai mā Rongo. Katoa ngā kai he mea

> whakahere ki a Rongo, ki te whetū anō ki a Matariki, arā, kia nui, kia tini tana hōmaitanga i te kai.[23]
>
> In ancient times before the people would begin planting, the sacred gardens were set-aside for the god of food, for Rongo. All the food within were offerings to Rongo, and to the star Matariki, so the produce for the year would be large and many.

This traditional Matariki practice was adapted by the Māori prophet and spiritual leader Te Kooti in 1879, and has since been embedded within both the huamata and pure ceremonies of the Ringatū faith.[24] This huamata ceremony is conducted on the first of June every year, and involves prayer and the planting of the small sacred garden. While the Ringatū church continues to maintain the huamata and pure rites, the actual origins of these practices are connected with Matariki. 'Although Ringatū do not overly celebrate the rising of Matariki, both the Te Huamata and Te Pure planting and harvesting rites that Te Ringatū adhere to, are deeply rooted in Maori cosmology and the ancient practice of star worship.'[25]

This traditional practice also gave rise to the proverb, 'Rewa ake i te ata, ko Matariki, ei ko te huamata.'[26] 'When Matariki rises in the morning, it is time to enact the huamata.'

A well-known proverb that connects Matariki to food is the saying 'Matariki, ahunga nui', meaning 'Matariki, provider of food'.[27] Generally, this is applied to the Matariki celebrations of winter when the storehouses and kūmara pits are full and people gather together to feast. Another saying is 'Matariki rahi nui', meaning 'Matariki the great slave'. 'Equivalent to a chief who collects all food of the country from different tribes'. This is according to the Waikato tohunga Paora Tūhua. Indeed, it may be said, 'Matariki rahi nui ko te ahua ia e', meaning 'the attributes of Matariki are boundless'.[28]

Matariki has a special association with the kūmara, the kererū and the korokoro. A group of four stars were observed in the winter mornings and forecast the productiveness of the kūmara gardens. These stars were Matariki, Tautoru (Orion's Belt), Puanga (Rigel in Orion) and Whakāhu (Castor and Pollex in Gemini). 'If their appearances indicated a favourable season, planting began in September; if otherwise the task was put off until October'.[29] Later in the year, during the month of Whiringa ā-nuku and Whiringa ā-rangi (approximately October and November), when Matariki was high in the night sky and the kūmara had sprouted and had grown to a certain height, food was hung in trees near the kūmara plantations. This ritual was observed to ensure the continued growth of the kūmara plant through spring and into summer. The association between Matariki and the kūmara was widespread throughout traditional Māori society, and Grey states that 'all tribes made offerings of their first sweet potatoes to Matariki'.[30] For its connection to the kūmara, Matariki is sometimes referred to as 'Hoko kūmara', another symbolic name for this cluster.

Of the food sources that were harvested during the time of Matariki, the most notable were the kererū and the korokoro. The kererū was harvested after Matariki appeared in the morning sky in Pipiri. Within the tribal boundary of Tūhoe, the kererū is a sacred bird and was an extremely important food source. Not only prized for its taste, the kererū was also used for trading with neighbouring tribes and as a food source to honour guests. Therefore, the kererū and its related activities were bound in prestige and ceremony.[31] It is recorded that at the rising of Matariki, tohunga would gather in a cave on Maungapōhatū. This cave is called Te Ana Whakatangi Whaititiri, meaning 'the thundering cave', and it was here that karakia would be offered to Matariki to open the kererū season. At the conclusion of the ceremony, thunder would be heard resounding in the sky, and this was a signal to the people of Maungapōhatū and Ruatāhuna that the season was open. As Best records:

> At Maungapohatu is a cave named Te Ana whakatangi whaitiri [the thunder-sounding cavern]. At that place the season for taking

Kūmara storage pit and plantation in Ōmaio, East Coast of the North Island. Photograph: Erica Sinclair 2015.

> products of the forest was opened by means of certain ceremonial performances, performed each year, and during which thunder was caused to resound, to give force to the rite.[32]

The last individual to conduct this ceremony within Tūhoe was Rawiri Te Kōkau in the 1930s.[33]

The proverbs 'kua kitea a Matariki, ā, kua maoka te hinu,' ('when Pleiades rises the fat is cooked') and 'ka rere a Matariki, ka wera te hinu' ('when the Pleiades appear, the fat is heated') refer to the kererū being captured, cooked and preserved in its own fat, during the time of Matariki.[34]

Korokoro were taken by Māori during the winter and early spring, as McDowall elaborates: 'What Māori clearly did understand very well was that adult lamprey leave the seas during late winter and spring and begin a long upstream migration to their spawning grounds'.[35]

This spawning cycle coincides with the sighting of Matariki, hence the saying, 'Ka kitea a Matariki ka rere te korokoro'; 'When Matariki is seen by the eye of man, then the korokoro migrates and is slain by the Maori people'.[36]

Apart from the pursuing of kererū and korokoro, there was little other food-related activity during Matariki. This is due to the pre-dawn rising of this cluster coinciding with the coldest and most unproductive time of year. Furthermore, Matariki rises after the harvest has finished, when the storehouses and kūmara pits are full with produce. It must be stressed that apart from the māra tautāne, Matariki was not a time for planting or harvesting crops. General planting occurred about October, and the harvest took place sometime around March or April, and was known to Māori as 'hauhakenga', meaning 'harvest'. For the most part Matariki was a time of pleasure, feasting and merrymaking.[37]

Opposite page: Painted rafter in Te Whai a te Motu meeting house in Ruatāhuna, showing Te Kurapa spearing kererū. This activity was undertaken after Matariki was seen in the sky. Photograph: Te Kirikatōkia Tukura Rangihau 2016.

A commonly misinterpreted proverb regarding Matariki is the saying 'Hauhake tū, ka tō Matariki', which is at times translated to mean 'the lifting of the crops begins when Pleiades sets'.[38] As we have seen, Matariki sets in Haratua, or around the month of May, after the crops have been harvested. Waiting until the setting of Matariki in the month of Haratua to harvest the gardens would be ill-advised, as it is early winter and the frost and cold weather would impact on the produce. The word 'tū' within this proverb means 'to end' or 'to cease'. Therefore, the true meaning of this saying is 'Matariki is setting so the harvest should be finished or quickly coming to an end.'

As recorded in *Te Waka Maori o Niu Tirani*: 'Haste with the harvest; Matariki (Pleiades) is setting (ie the season is advanced).'[39] The meaning of this proverb is that the labours of late summer and early autumn have now finished and a time of celebration and relaxation is about to begin.

THE WEATHER OF MATARIKI

Aside from food, Matariki has a powerful relationship with the weather. As already explained, the stars Waipunarangi and Ururangi within the Matariki cluster are connected to the rain and the wind, and the appearance of these stars in the morning sky in Pipiri would forecast the weather for the impending year. The connection between Matariki and weather is reaffirmed within a number of Māori proverbs and phrases. For example, the saying 'ka rere ngā purapura o Matariki', meaning 'the seeds of Pleiades are falling', refers to snow falling in winter. Likewise, the statement 'Matariki tāpuapua' is used to describe rain puddles that cover the ground in the cooler months.[40] The expression 'ngā taritari o Matariki' is applied when describing the chilling winter winds. This phrase is noted in a pātere[41] recorded by Ngata:

> Mō te makariri, e totope nei te hukarere,
> Ngā taritari o Matariki

> In the winter time, heralded now by snowstorms
> And this cold weather of Matariki[42]

Matariki has a relationship to both the winter and summer seasons, and this is seen in the phrase 'te paki o Matariki', denoting the calm fine weather of summer. Riley expands on the bond between Matariki and the summer period:

> When Rata had built the Tainui he was advised the canoe should leave Hawaiki for Aotearoa in the month of Akaaka-nui (December) when Matariki, the Pleiades star constellation, was above, with its calm, fair weather.[43]

Te paki o Matariki has a special meaning for the people of the Tainui canoe and those associated with the Kīngitanga movement. Matutaera Tāwhiao, second Māori king, took Matariki, among other symbols, as an emblem and coat of arms for his people. Te paki o Matariki, an expression promising fine weather and good fortune, was taken by Tawhiao to symbolise peace and calm in New Zealand. *Te Paki o Matariki* was also the name given to the newspaper of the Kīngitanga.[44]

MATARIKI AND NAVIGATION

Matariki was among a select group of stars that was observed by early navigators when travelling throughout the islands of Polynesia. The knowledge that these explorers had of their environment and the night sky was remarkable, and on the open ocean their wisdom would have been a crew's salvation. A number of famous voyaging canoes used the Matariki cluster when undertaking their journey from central Polynesia to Aotearoa,[45] and this knowledge is said to have descended from Kupe, the first to make this epic voyage. When Kupe was asked for directions to Aotearoa he is said to have replied,

Carved door at Turongo house, Ngaruawahia, showing Te Paki o Matariki, the Māori king's coat of arms.
Godber, Albert Percy, 1875–1949: Collection of albums, prints and negatives.
Ref: APG-1501-1/4-G. Alexander Turnbull Library, Wellington, New Zealand.

Hoturoa Kerr at Kennedy Bay standing in front of the waka Haunui.
Photograph: Taputu Raea 2016.

'Let it be directed to the left of the rising sun and until it is well up the heavens, and so continue until the Pleiades rise above the ocean surge, that you may reach land'.[46]

The importance of Matariki as a navigation tool was noted by Best:

> Writing of stars in his Polynesian Researches Ellis says: 'These were their only guides in steering their fragile barks across the deep. When setting out on a voyage some particular star or constellation was selected as their guide during the night ... The Pleiades were a favourite guiding-star with these sailors, and by them, in the present voyage, we steered during the night'.[47]

Even in a modern context, Matariki continues to guide a new generation of Māori and Polynesian navigators across the Pacific Ocean. Traditional navigational expert and sailor Hoturoa Kerr is part of the renaissance associated with Polynesian ocean voyaging. He states that the observation of Matariki remains to this day a sign to inform the navigator that the sailing season has begun, and it is time to leave central Polynesia and journey to Aotearoa.[48]

MODERN MATARIKI

With the mass arrival of European settlers to Aotearoa in the latter part of the nineteenth century, and the introduction of new technologies, religions, education and political systems, many traditional Māori practices disappeared. The ravages of new illnesses, war and land alienation all added to the erosion of Māori culture,[1] and for the most part Matariki was discontinued as a practice. Some aspects associated with Matariki, such as the huamata and pure, were encompassed within new Māori religious movements like the Ringatū faith. At the same time, various tribal experts across the country maintained regional specific knowledge about Matariki. However, the elaborate Matariki celebrations of the ancient Māori world had all but disappeared by the beginning of the twentieth century.

Then in the early 1990s, Matariki underwent a revival spearheaded by interested Māori groups and iwi such as Ngāti Kahungunu in the Hawkes Bay.[2] A new celebration was then founded in Wellington and was hosted by the Museum of New Zealand Te Papa Tongarewa, which included a number of high-profile events with public lectures, cultural performances and the viewing of Matariki in the pre-dawn sky. This in turn ignited widespread interest in this celebration, and since that time, Matariki has become part of the national conscience.

Advertising Matariki celebrations at Te Papa Tongarewa.
Photograph: Ann Hardy 2015.

Today Matariki has become a well-known and extensively celebrated event, not just in Māori communities but New Zealand-wide. Matariki is now being embraced by non-Māori, and it has found its way into schools and regional and city councils' events. Recently, it has been part of a commercial movement that takes place during the winter solstice. There has even been a proposal to solidify Matariki as a public holiday, through a Members' Bill that was tabled by the Māori Party in the House in 2009.[3]

In 2016, Te Wānanga o Aotearoa embarked on a month-long Matariki promotional tour to stimulate discussion around Matariki and the Māori New Year. Based on the research conducted for this publication, the tour presented a new vision for Matariki, focused on Māori traditional practices connected to Māori star lore. The tour was called 'Te Iwa o Matariki',[4] 'the nine of Matariki', in reference to research suggesting that some Māori maintain nine names for the different stars in Pleiades.

From left to right, Hika Taewa, Ngahuia Kopa, Paraone Gloyne, Kahurangi Maxwell and Dan Moeke facilitating the Te Iwa o Matariki road show 2016.
Photograph: Erica Sinclair 2016.

Matariki has been extensively revitalised in the past three decades and its growth will continue as this celebration becomes repurposed and reapplied for a new generation. Its growing importance as a symbol of both Māori and New Zealand identity is helping to solidify its position within mainstream society. The general population's knowledge and understanding of Matariki is increasing, and the many public Matariki celebrations and events have given this festival a strong visual presence throughout Aotearoa. The next stage in the evolution of Matariki must be in securing its future, nurturing its growth and maintaining its integrity.

THE FUTURE OF MATARIKI

During the month of June a host of Matariki activities take place throughout the country, including public lectures, balls, dinners, art exhibitions and the viewing of the cluster in the early morning sky. However, there remains some doubt as to the purpose of Matariki and its role within our modern world. For many there is a disconnect between what Matariki was, in a traditional sense, and what it is becoming, in a modern context. Currently questions exist around exactly when and how Matariki should be celebrated and who should be taking the lead in its revitalisation and future.

Perhaps to best understand what Matariki could and should be, we need to look beyond ourselves and examine Matariki activities in other parts of Polynesia. A relevant example is the Makahiki festival of Hawaiʻi, which was reintroduced in the early 1980s. Every year Hawaiʻian language and cultural advocates gather on the island of Kahoʻolawe to observe the rising and setting of Makali'i and to offer food to the god Lono, as part of an elaborate ceremony honouring traditional Hawaiʻian spirituality.[5] The practices involved are strikingly similar to the Māori records associated with Matariki. Furthermore, the Hawaiʻian Makahiki practices are more than just a celebration. They are a series of spiritual and religious ceremonies that reaffirm the belief and commitment the native Hawaiʻian people have to their gods, their language, their culture and their environment.

After more than twenty years of research and work within the Māori astronomy arena, I believe it is now both timely and appropriate for the traditional practice of Matariki to be resurrected, reintroduced and repurposed, in order for it to play a meaningful role in the lives of those who choose to be part of this movement. The Matariki ceremony needs to be repositioned again at the heart of Māori spirituality, and the setting and rising of this star cluster should be correctly observed and celebrated on a yearly basis. Pivotal to this movement is adherence to traditional Māori spirituality and the honouring of the different Māori gods. Hence, for this movement to be truly purposeful it must include practices such as whāngai hau and incantations to the various stars and Māori deity, and these practices need to be embedded in a ceremony that becomes the focal point for the celebration of Matariki. Importantly, this revitalised celebration needs to be religious, spiritual, cultural, environmental as well as social. Matariki needs to move beyond the assortment of social gatherings that have become normal practice during the winter month, and take on a more holistic approach. Matariki needs to be honoured, celebrated and replenished every year to ensure that those who have died since its last rising become stars in the night sky, and in order to secure the bounty of the impending season.

Ultimately these ceremonies, practices and the entire Matariki celebration will fail to have any real meaning unless there is a commitment by practitioners to reintroduce the ceremony, supported by Māori and wider communities, to celebrate the traditions of Matariki, and most importantly, that all involved in this movement have faith in the traditional beliefs and practices of Māori. If these various elements can come together as part of future Matariki celebrations, then this annual event can once again become central in the lives of people in Aotearoa.

A new Matariki movement centred on traditional practice must be underpinned by Māori culture and Māori language, for it is within the language and upon the culture that Matariki was established. Matariki has evolved in the past few decades to be an inclusive movement that has encompassed many peoples from a host of backgrounds. However, the future of this movement lies in its past, and the key to this past are those who have traditional knowledge and those who are practitioners and exponents of Māori language and culture.

CONCLUSION

Pleiades is recognised and celebrated by many cultures throughout the world, and the sighting of this cluster in the sky has many meanings. It is connected to life and death, planting and harvesting, summer and winter, fishing and navigation. It is a symbol of cultural identity for communities, tribes and nations across the globe. In Aotearoa, Matariki signalled the changing of the year, and its pre-dawn rising heralded in a period of remembrance, reflection, hope, preparation and celebration.

The observance of Matariki, like many cultural practices, suffered greatly due to the oppressive weight of colonisation, and by the middle of the twentieth century the adherence to Matariki was virtually non-existent. While some aspects of the Matariki practice became incorporated within various Māori religious movements and some cultural experts maintained knowledge around the traditional ceremony, the importance of Matariki as a religious, cultural and environmental symbol had all but disappeared. Then, in the early 1990s a movement was established to revitalise the Matariki celebration. Initially led by Te Papa Tongarewa, Matariki began to gain traction and grow as a movement that spread across the country. Today it is celebrated in various ways and at different times, and continues to entrench its position as an important yearly festival within Aotearoa.

However, Matariki is more than a cluster of stars that marks the changing of the season and the winter solstice. It is more than an environmental indicator that predicts the new season's growth, and it is more than a symbol of unity, togetherness and hope. Matariki is greater than its connection to new life and its remembrance of the deceased. Matariki transcends boundary, religion, political agenda and even race. Matariki has different meanings for different people, and in a new age it has become a marker, not only of culture but also of national identity.

The eyes of the god Tāwhirimātea were thrown into the sky in a spiteful fit of anger and a sorrow-filled act of love. These stars served a purpose for the ancestors of the Māori, and in this modern society they have been revived, taking on new meaning for a new generation. Perhaps the future of Matariki is best portrayed in the following proverb: 'Matariki ki tua o ngā whetū'; 'Matariki of endless possibilities'.[6]

Matariki over Aoraki in the South Island.
Photograph: Kōmene Cassidy 2015.

MATARIKI PROVERBS

The following list is a collection of proverbs connected to Matariki. Included with each is a description of its meaning and the context in which the phrase can be applied.

'Hauhake tū, ka tō Matariki.'[1]
'The harvest ends when Matariki sets.'

Matariki sets in the western sky with the sun during the month of Haratua. This is a sign that the harvest season has come to an end. The saying is used before the onset of winter, to ensure people have prepared themselves for the cold months to follow.

'Ka kitea a Matariki, ka rere te korokoro.'[2]
'When Matariki is seen, the lamprey migrates.'

The migration of the lamprey coincides with the pre-dawn rising of Matariki in the winter sky. This proverb celebrates this ecological event.

'Ka mahi ngā kanohi tīkonga a Matariki.'[3]
'The ever-alert and protruding eyes of Matariki.'

This proverb is said of a person who is wakeful and alert at night. The origin of this saying is the story in which Tāwhirimātea plucked out his eyes and threw them into the heavens, where they became Matariki. These protruding eyes are seen unblinking in the heavens, staring intently at the earth. A similar saying is 'Tēnā ngā kanohi kua tīkona e Matariki'.[4]

'Ka rere a Matariki, ka wera te hinu.'[5]
'When Matariki rises, the fat is heated.'

Traditionally the rise of Matariki was a time when kererū were taken in large numbers. Once caught, these birds were preserved in their own fat and stored to be eaten at a later date. The heating of fat to preserve this Māori delicacy occurs during the time of Matariki and is the origin of this saying. Similar versions of this saying are 'Ka kitea a Matariki, nā kua maoka te hinu'[6] and 'Ka puta a Matariki, kua maoke te hinu'.[7]

'Ka rere ngā purapura a Matariki.'[8]
'The seeds of Matariki are falling.'

This comment refers to the snowfalls of winter when Matariki is in the morning sky.

'Ka puta Matariki, ka rere Whānui, ko te tohu o te tau.'[9]
'Matariki appears as Whānui flees; this is the sign of the New Year.'

Whānui is the star Vega in the constellation of Lyra. Whānui sets in the northern sky in the early morning of winter just before the rising of Matariki in the eastern sky. This event is a sign of the New Year and is recorded in this proverb.

'Ko Matariki te kaitō i te hunga pakeke ki te pō.'[10]
'Matariki draws the frail into the endless night.'

Matariki has a strong association with the dead, and during its rising in winter the old and weak often pass away. This statement reflects this occurrence.

'Kua haehae ngā hihi o Matariki.'[11]
'The rays of Matariki are spread.'

This comment is spoken when Matariki is seen bright in the night sky, and the rays of its various stars carry messages for the people.

'Matariki ahunga nui'.[12]
'The great mounds of Matariki'.

Matariki is a time when food like kūmara is stored in large piles and shared among friends, family and the wider community. This proverb speaks of hospitality, togetherness and the sharing of food.

'Matariki hunga nui.'[13]
'Matariki of many people.'

During Matariki festivities people gather together to celebrate the passing of one year and the hope of a new one. The gathering of people is recorded in this phrase.

'Matariki kāinga kore.'[14]
'Homeless Matariki'.

Matariki is always on the move, accompanying the dead as they travel the sky. This cluster is also somewhat isolated, with no other bright stars located near its vicinity. For these reasons Matariki is said to be without a home. This proverb can be applied to a person who is constantly travelling, or to someone who isolates themselves from others.

'Matariki kanohi iti.'[15]
'The detailed eye of Matariki.'

This proverb is sometimes translated to mean 'Matariki of little hospitality' because some believe there is little food during this season.[16] However, as we have seen, Matariki happens after the harvest when the food stores are full, and this is a period of great feasting and sharing. The above proverb is actually connected to people who pay attention to small details in search of perfection. This is similar to the phrase 'ngao matariki', which means 'working timber with a small adze to accomplish a fine finish'.[17]

'Matariki tāpuapua.'[18]
'The pools of Matariki.'

The rising of Matariki in winter is synonymous with rain; the pooling of rain water on the ground in winter is immortalised in this saying.

'Nā Matariki te karaka tono, kia korokoro. Ka tīmata te kake i ngā awa ki te whānau i o rātou nei tamariki.'
'When Matariki calls, the lamprey begin their ascent of the river ways to give birth to their young.'

The rising of Matariki is connected to the spawning of the lamprey.

'Ngā kai a Matariki nāna i ao ake nei.'[19]
'Food that is scooped up by Matariki.'

Matariki is connected with food, and the nature of its early morning appearance in winter is said to determine the bounty of the impending year. The above statement is said of Matariki when it rises in the month of Pipiri and food is cooked and offered upon its appearance. Matariki is weak and cold from carrying out its yearly duty, and Māori would cook food to replenish its strength. This proverb is said during the early morning New Year ceremony.

'Te ope o te rua Matariki.'[20]
'The company from the cavern of Matariki.'

Matariki has a strong connection to the dead, carrying the deceased across the sky night after night. This proverb can be used when speaking about the many illustrious chiefs who have departed the world of the living and have gathered in the cavern of Matariki. Alternatively, this proverb can be used when addressing a company of leaders and experts, likening them to the many great chiefs who have passed into the afterlife.

'Matariki whanaunga kore; Matariki tohu mate.'[21]
'Matariki the kinless; Matariki a sign of death.'

Matariki rises in winter, a time when the weak and frail often succumb to illness and die. Matariki is often seen in the sky when people pass away and this statement is a reminder to people that while Matariki is a time of celebration and enjoyment, it also has a connection to death.

ENDNOTES

ACKNOWLEDGEMENTS

1. This section of karakia (traditional incantation) was given by Pou Temara, 2013.
2. This translation and others throughout the book are by the author, unless stated otherwise.

MATARIKI – THE STAR OF THE YEAR

1. Matariki is at times referred to as 'te whetū o te tau', the star of the year, or 'te whetū tapu o te tau,' the venerated star of the year. The term 'te whetū o te tau' is recorded in a number of sources. See *Te Whetu o Te Tau*. (Vol. 1, No. 1, Akarana, 1 Hune 1858), https://paperspast.natlib.govt.nz/newspapers/WHETU18580601.2.5?query=Te%20Whetu%20o%20te%20tau.%20Akarana.
2. E. Best, *The Maori as he was: a brief account of Maori life as it was in pre-European days* (Wellington: Government Printer 1952) 135.
3. E. Tregear, *The Maori Race* (Whanganui: A. D. Willis Printer and Publisher 1904) 22.
4. E. Best, *Tuhoe: Children of the Mist* (Auckland: Reed 1996) 780–858.
5. E. Best, *The Astronomical Knowledge of the Maori* (Wellington: Government Printer 1955) 64.
6. H. Tuahine, *Te Tāhū o Ranginui: Whakatūria te Whare Kōkōrangi*, unpublished master's thesis, University of Waikato, 2015.

MATARIKI/PLEIADES

1. In 1771, French astronomer and comet-hunter Charles Messier compiled a list of celestial objects to help him identify bodies that were not comets.

This became the Messier catalogue, of which Pleiades or Matariki is the forty-fifth object on the list, hence the name M45.

2. Apparent magnitude is a scale that astronomers use to measure the brightness of an object in the night sky as seen from the Earth. The lower scoring the object the brighter it is. For instance the Sun has an apparent magnitude of -27 and the full moon -13. Venus the brightest planet measures -5 and the brightest star Sirius is -1.5.
3. E. Best, 1952, 185.
4. A. C. Gifford, *In Starry Skies, the Solar System.* (Wellington: New Zealand Astronomical Society 1937) 8.
5. E. Best, *The Maori.* Vol. 2 (Wellington: Harry H. Tombs 1924) 212.
6. M. Rappengluck, 'The Pleiades in the "Salle des Taureaux" grotte de Lascaux. 'Does a rock picture in the cave of Lascaux show the open star cluster of the Pleiades at the Magdalénien era ca. 15.300 BC?', in *IV SEAC Meeting, Astronomy and Culture* (1997) 217–225.
7. R. H. Allen, *Star Names, their Lore and Meaning* (New York: Dover 1963) 392.
8. F. & K. Wood, *Homer's Secret Odyssey* (London: The History Press 2014).
9. F. Joseph, *The Atlantis Encyclopedia* (New Jersey: New Page Books 2005) 226.
10. P. Bogard, *Let There Be Night – Testimony on behalf of the dark* (Reno: University of Nevada Press 2007).
11. R. Hard, *Eratosthenes and Hyginus Constellation myths with Aratus's Phaenonena* (United Kingdom: Oxford University Press 2015) 92.
12. M. Andrews, *The Seven Sisters of the Pleiades: Stories from around the world* (Melbourne: Spinifex Press 2004).
13. W. T. Olcott, *Star Lore. Myths, Legends and Facts* (New York: Dover Publications 2004) 415.
14. A. Johns, *Baba Yaga – The Ambiguous Mother and Witch of the Russian Folktale* (New York: Peter Lang Publishing 2014) 9.
15. Llewellyn, *Llewellyn's 2013 Sabbats Almanac. Samhain 2012 to Mabon 2013* (Woodburry: Llewellyn Worldwide Ltd 2012) 102.
16. M. Andrews, 2004, 325.
17. B. Brady, *Brady's Book of Fixed Stars* (Boston: Weiser Books 1998) 237–238.
18. W. Horowitz, *Mesopotamian Cosmic Geography* (Indiana: Eisenbrauns 1998).
19. K. G. Gopakumar, *The Great Year and Virgin Comets* (India: Gopa Kumar 2013).
20. M. Andrews, 2004, 26.
21. S. Renshaw & S. Ihara, 'A Cultural History of Astronomy in Japan', in H. Selin, *Astronomy Across Cultures: The history of non-western astronomy* (Berlin: Springer Science+Business Media 2000) 390.

22. R. Burnham, *Burnham's Celestial Handbook: An Observer's Guide to the Universe Beyond the Solar System* (New York: Dover Publications 1978) Vol III, 1864.
23. D. Harness, *The Nakshatras. The Lunar Mansions of Vedic Astrology* (Delhi: Motilal Banarsidass Publishers 2000) 11.
24. N. Red Star, *Star Ancestors: Extraterrestrial Contact in the Native American Tradition* (Rochester: Bear & Company 2000) 68.
25. R. Hail, *Cherokee Astrology: Animal Medicine in the Stars* (Rochester: Bear & Company 2000) 9.
26. N. Maryboy & D. Begay, *Sharing The Skies: Navajo Astronomy* (Tucson: Rio Nuevo Publishers 2010) 40–41.
27. N. Red Star, 2000, 89.
28. P. R. Steele, *Handbook of Inca Mythology* (Santa Barbara: ABC-CLIO 2004) 144.
29. D. Johnson, *Night Skies of Aboriginal Australia* (Sydney: Sydney University Press 2006) 74.
30. T. Andrew, *Dictionary of Nature Myths: Legends of the Earth, Sea and Sky* (New York: Oxford University Press 1998) 76.
31. R. Johnson, J. Mahelona & C. Ruggles, *Nā Inoa Hōkū. Hawaiian and Pacific Star Names* (Bognor Regis: Ocarina Books 2015) 190–91.
32. M. Beckwith, *The Kumulipo: A Hawaiian Creation Chant* (Honolulu: University of Hawaii Press 1981) 121.
33. W. Levin, R. Reeve, F. Salmoiraghi & D. Ulrich, *Kaho'olawe, Nā Leo o Kanaloa: Chants and stories of Kaho'olawe* (Hawai'i: Ai Pōhaku Press 1995) 88–92.
34. G. Turner, *Samoa, A Hundred Years Ago and Long Before* (London: Macmillan 1884) 202.
35. M. Andrews, 2004, 321.
36. D. Tyerman & G. Bennet, *Voyages and Travels Round the World* (London: John Snow 1841); T. Henry, 'Tahitian Astronomy', in *Journal of Polynesian Society* (Vol. 16. No. 2) 101–104.
37. G. Magli, *Mysteries and Discoveries of Archaeoastronomy: From Giza to Easter Island* (New York: Praxis Publishing 2005) 250–251.
38. J. Sissons, *The Polynesian Iconoclasm: Religious Revolution and the Seasonality of Power* (New York: Berghahn Books 2014) 26.
39. M. Makemson, *The Morning Star Rises: An Account of Polynesian Astronomy* (New Haven: Yale University Press 1941) 230.
40. E. Beaglehole, *Ethnology of Pukapuka* (Honolulu: Bernice P. Bishop Museum 1938).
41. E. S. Handy, *The Native Culture in the Marquesas*. Bulletin 48 (Honolulu: Bernice. P. Bishop Museum 1923).

THE MEANING OF MATARIKI

1. E. Best, *Tuhoe: Children of the Mist* (Auckland: Reed 1996) 812.
2. M. R. Orbell & G. Moon, *The Natural World of the Maori* (Auckland: David Bateman 1985) 69.
3. M. Beckwith, *Hawaiian Mythology* (Honolulu: University of Hawai'i Press 1976) 368; P. Buck, *Ethnology of Mangareva.* Bulletin 157 (Honolulu: Bernice P. Bishop Museum 1938).
4. J. White, *Ancient History of the Maori, His Mythology and Traditions Horouta or Takitumu Migration.* Vol. I (Wellington: Government Printer 1887) 20; M. Makemson, *The Morning Star Rises. An Account of Polynesian Astronomy* (New Haven: Yale University Press 1941) 76.
5. K. Clark, *Maori Tales & Legends* (London: David Nutt 1896); J. Tobin, *Stories from the Marshall Island* (Honolulu: University of Hawai'i Press 2002).
6. Te Kōkau manuscript (1898–1933).
7. Ranginui and Papatūānuku, also referred to as the sky father and earth mother, are acknowledged by Māori as the origins of the universe. Between the dark space that existed between them lived the Māori gods. It was Tāne, god of the forest, who forced them apart, ultimately creating the world we know today.
8. Self-inflicted wounds, lacerations and the cutting of hair were customs traditionally performed by Māori expressing grief during the mourning process. This originates with Tāwhirimātea who removed his own eyes.
9. J. White, *Ancient History of the Maori, His Mythology and Traditions Horouta or Takitumu Migration.* Vol. I (Wellington: Government Printer 1887) 15.
10. P. Temara, personal communication, 15 March 2014.
11. T. H. Te Rangi, 'Tera Matariki', in *The Maori Messenger – Te Karere Maori* (Vol. 7, No. 18, 30 November 1860) 39.
12. E. Best, *The Astronomical Knowledge of the Maori* (Wellington: Government Printer 1955) 40, 56; A. Ngata & P. T. H. Jones, *Ngā mōteatea: He maramara rere no nga waka maha.* Vol. 3 (Wellington: Polynesian Society 2004b) 550–555.
13. E. Tregear, *The Maori Race* (Wanganui: A. D. Willis Printer and Publisher 1904) 401; E. Best, 1996, 813.
14. P. Smith, 'Wars of the northern against the southern tribes of New Zealand in the nineteenth century Part III', in *Journal of the Polynesian Society* (Vol. 9, No. 1, 1900) 20; E. Best, 'Maori star names', in *Journal of the Polynesian Society* (Vol. 19. No. 2, 1910) 97.
15. E. Best, *The Astronomical Knowledge of the Maori* (Wellington: Government Printer 1955) 52.
16. Te Kōkau manuscript (1898–1933); H. Mead, *Nga Taonga Tuku Iho a Ngati Awa: Ko nga tuhituhi a Hamiora Pio, Te Teko (1885-1887)* (Wellington: Department of Māori, Victoria University of Wellington 1981) 45; E. Best, 1996, 813.
17. E. Best, 1955, 34.

18. E. Krupp, *Beyond the Blue Horizon: Myth and Legends of the Sun, Moon, Stars and Planets* (New York: HarperCollins 1991) 247.
19. K. G. Gopakumar, *The Great Year and Virgin Comets* (India: Gopa Kumar 2013) 65.
20. Turongo House, *Te Tumu Kōrero* (Ngaruawahia: The House 1983) 47; K. Leather & R. Hall, *Tātai Arorangi: Māori Astronomy. Work of the Gods* (Paraparaumu: Viking Sevenseas 2004) 63.
21. T. Rolleston-Cummins, *The Seven Stars of Matariki* (Wellington: Huia 2008); R. Hard, *The Routledge Handbook of Greek Mythology* (New York: Routledge 2004) 518.
22. Te Kōkau manuscript (1898–1933).
23. E. Tregear, *The Maori Race* (Wanganui: A. D. Willis Printer and Publisher 1904) 22.
24. M. Riley, *Wise Words of the Māori: Revealing history and traditions* (Paraparaumu: Viking Sevenseas 2013) 661.
25. W. Reeves, *The Long White Cloud Aotearoa* (London: Horace Marshall & Son 1898) 60.
26. Pipiri is the first month of the Māori year, which is June in the Gregorian calendar. It is important to note that the Māori calendar follows the lunar cycle as opposed to a numerical system like the Western calendar. This point is important when trying to identify the Māori New Year and the rising of Matariki.
27. T. E. Doone, *The Maori Past and Present* (London: Seeley Service and Co. 1927) 70; B. Mitcalfe, 'Te Rerenga Wairua – Leaping Place of the Spirits', in *Te Ao Hou: The New World* (June 1961) 43.
28. Ministry for Culture and Heritage, *Te Taiao – Māori and the Natural World* (Auckland: David Bateman 2010) 24.
29. H. W. Williams, *Dictionary of the Māori Language* (Wellington: Legislation Direct 2003) 457, 422, 225.
30. H. M. Mead & N. Grove, *Ngā Pēpeha a ngā Tīpuna* (Wellington: Victoria University Press 2003) 61.
31. Kererū is the native New Zealand pigeon, or wood pigeon *Hemiphaga novaeseelandiae*. It is also known to Māori as kūkupa and kūkū; M. Riley, *Māori Bird Lore* (Paraparaumu: Viking Sevenseas 2001) 26; Ministry for Culture and Heritage, 2010, 23.
32. W. H. Williams, 2003, 477.
33. M. Riley, 2013, 348; additional Māori names for korokoro include kanakana (South Island) and piharau; R. M. McDowall, *Ikawai: Freshwater fishes in Māori culture and economy* (Christchurch: Canterbury University Press 2011) 123.
34. E. Best, *Maori Religion and Mythology*. Part 2 (Wellington: Te Papa Press 2005b) 393.
35. M. Riley, 2001, 26.
36. W. H. Williams, 2003, 469.
37. E. Best, 1996, 781.
38. Te Kōkau manuscript (1898–1933).

MATARIKI NEW YEAR

1. T. Maxwell, *Te Kōpura*. Unpublished master's thesis, University of Waikato, New Zealand, 1998, 34.
2. E. Best, *The Maori Division of Time* (Wellington: A. R. Shearer, Government Printer 1973); Ministry for Culture and Heritage, *Te Taiao – Māori and the Natural World* (Auckland: David Bateman 2010).
3. M. Roberts, F. Weko & L. Clarke, *Maramataka: the Māori Moon Calendar*. Research Report No. 283 (Christchurch: Lincoln University 2006); W. Tāwhai, *Living by the Moon: Te Maramataka a Te Whānau ā Apanui* (Wellington: Huia 2013) 13.
4. C. Ruggles, *Ancient Astronomy: An Encyclopedia of Cosmologies and Myth* (Santa Barbara: ABC-CLIO 2005) xxi.
5. E. Best, *Some Aspects of Maori Myth and Religion* (Wellington: Government Printers 1954) 8.
6. E. Best, 1973, 19.
7. H. Mahupuku, *Whakapapa Tuupuna* (MS, private collection, Whakatāne 1854. Cited by P. Hohepa in his translation document (1992)).
8. Developed by Pope Gregory XIII in 1582, the Gregorian calendar, also known as the Christian or the Western calender, has become the most accepted and widely used calender in the world.
9. J. Kien, *The Battle Between the Moon and Sun* (Boca Raton: Universal Publishers 2003) 140; E.W. Hetherington & N. S Norriss, *Astronomy and Culture* (Santa Barbara: Greenwood Press 2009) 78.
10. H. Harris, R. Matamua, T. Smith, H. Kerr & T. Waaka, 'A review of Māori Astronomy in Aotearoa-New Zealand', in *Journal of Astronomical History and Heritage* (Vol. 16, No. 3, 2013) 325–336: 330; M. Roberts et al, 2006, 7–13.
11. M. Roberts et al. 2006, 16.
12. E. Best, *The Maori as he was: a brief account of Maori life as it was in pre-European days* (Wellington: Government Printer 1952) 50.
13. E. Best, 'Maori Star Names', in *Journal of Polynesian Society* (Vol. 19, No. 2, 1910) 10–11; E. Best, *Tuhoe: Children of the Mist* (Auckland: Reed 1996) 789.
14. E. Best, 1973, 49; 'Honorific Terms, Sacerdotal Expressions, Personifications, etc., Met with in Maori Narrative' in *Journal of the Polynesian Society* (Vol. 35, No. 140, 1926): 333.
15. Ministry for Culture and Heritage, 2010, 23.
16. Te Taura Whiri i te Reo Māori, *Matariki* (Wellington: Te Taura Whiri i te Reo Māori 2010) 8.
17. T. Ruka. 'Reta ki te Etita', *Pipiwharauroa He Kupu Whakamarama* (No. 14, 1 April 1899) 4.
18. J. Batten, *Celebrating the Southern Seasons: Rituals for Aotearoa* (Auckland: Random House 2005) 59; Ministry for Culture and Heritage, 2010, 22.

19. H. Harris, R. Matamua, T. Smith, H. Kerr & T. Waaka, 'A review of Māori Astronomy in Aotearoa-New Zealand', in *Journal of Astronomical History and Heritage* (Vol. 16, No. 3, 2013) 330.
20. W. Tāwhai, 2013, 41.
21. E. Best, *Maori Religion and Mythology*. Part 1 (Wellington: Te Papa Tongarewa 2005) 76–77.
22. W. Tawhai, 'Te Maramataka Māori', *Waka Huia* (TVNZ 13 June 2009); Te Kōkau manuscript (1898–1933).
23. 'Te Aroha o Rangi-nui Kia Papatuanuku', *Te Toa Takitini* (No. 11, 1 June 1922) 10.
24. Department of Internal Affairs, *Ngā Tāngata Taumata Rau* (Wellington: Bridget Williams Books & Department of Internal Affairs, 1994) 115–119; A. Taonui, 'Ki Te Kai Ta o Te Wananga', *Te Wananga* (Vol. 2, 21 August 1875) 163–164.
25. E. Best, 1973, 15.
26. L. Hakaraia, *Matariki The Māori New Year* (Auckland: Reed 2004) 28–29.
27. A. Ngata & H. M. Mead, *Ngā mōteatea: He maramara rere no nga waka maha*. Vol. 4 (Wellington: Polynesian Society 2004) 72–75.
28. G. Grey, *Ko nga Moteatea me nga Hakirara o nga Maori* (Wellington: Stokes 2010) 346.
29. E. Best, 1996, 811–812.
30. Ibid, 812.
31. A. Hughes, *Matariki: Everything there is to know about it* (6 June 2016) http://www.stuff.co.nz/science/80739387/Matariki-Everything-there-is-to-know-about-it
32. There are many variations of the Māori lunar calendar, depending on tribe, region and local environment. This particular version is common within the Mataatua district; however, many shown are about the same throughout the country.

RISING AND SETTING OF MATARIKI

1. J. Evans, *The History and Practice of Ancient Astronomy* (New York: Oxford University Press 1998).
2. This image is an adaption from an earlier picture by Freedom Cole. See: www.svatantranatha.blogspot.com.au/2014/09/heliacal-cycle.html
3. J. White, *Ancient History of the Maori, His Mythology and Traditions Tai-Nui*. Vol. IV (Wellington: Government Printer 1888a) 131.
4. S. Webb, *Measuring the Universe: The Cosmological Distance Ladder* (Leicestershire: Springer-Praxis 2001) 73.
5. L. V. Jones, *Stars and Galaxies* (Santa Barbara: Greenwood Press 2010) 41.

6. T. M. Leaman, D. W. Hamacher & M. T. Carter, 'Aboriginal Astronomical Traditions from Ooldea, South Australia', Part 2: Animals in the Ooldean Sky, in *Journal of Astronomical History and Heritage* (Vol. 19, No. 1, 2011, Preprint) 11.
7. 'Te Aroha o Rangi-nui Kia Papatuanuku', *Te Toa Takitini* (No. 10, 1 May 1922) 13.
8. A. Ngata & H. M. Mead, *Ngā mōteatea: He maramara rere no nga waka maha.* Vol. 4 (Wellington: Polynesian Society 2004) 262–63.
9. Te Taura Whiri i te Reo Māori, *Matariki: Aotearoa Pacific New Year* (Wellington: Te Taura Whiri i te Reo Māori 2005) 3.
10. E. Best, *The Astronomical Knowledge of the Maori* (Wellington: Government Printer 1955) 53.
11. M. Riley, *Māori Bird Lore* (Paraparaumu: Viking Sevenseas 2001) 24.
12. Te Kōkau manuscript (1898–1933).
13. J. White, *Ancient History of the Maori, His Mythology and Traditions Tai-Nui.* Vol. V. (Wellington: Government Printer 1888b) 97.
14. J. White, *Ancient History of the Maori, His Mythology and Traditions Horouta or Takitumu Migration.* Vol. I (Wellington: Government Printer 1887) 15.
15. H. Kereopa, 'Matariki', *Waka Huia* (TVNZ 2001).
16. Te Kōkau manuscript (1898–1933).
17. M. R. Orbell & G. Moon, *The Natural World of the Maori* (Auckland: David Bateman 1985) 70; E. Best, 1955, 6.
18. E. Best, 1955, 41.
19. A. Ngata & P. T. H. Jones, *Ngā mōteatea: He maramara rere no nga waka maha.* Vol. 3 (Wellington: Polynesian Society: 2004b) 560–63.
20. H. M. Mead & N. Grove, *Ngā Pēpeha a ngā Tīpuna* (Wellington: Victoria University Press 2003) 383.
21. A. Ngata & H. M. Mead, 2004, 73–75.
22. C. Kirkwood, *Koroki, my King* (Ngaruawahia: Turongo House 1999) 22.
23. H. M. Mead, *Tikanga Māori. Living by Māori Values* (Wellington: Huia 2003) 145.
24. Radio New Zealand, *Whaikoorero: Ceremonial Farewells to the Dead* (Wellington: Continuing Education Unit, Radio New Zealand 1981) 22.
25. A. Ngata & H. M. Jones, *Ngā mōteatea: He maramara rere no nga waka maha.* Vol 2 (Wellington: Polynesian Society 2004a) 86–87.

MATARIKI: WHĀNGAI I TE HAUTAPU

1. H. W. Williams, *Dictionary of the Māori Language* (Wellington: Legislation Direct 2003) 488.
2. Tohunga, 'The Wisdom of the Maori' *The New Zealand Railway Magazine* (Vol. 10, No. 7, 1935) 45.
3. H. Dansey, 'Matariki', in *Te Ao Hou: The New World* (Vol. 61, 1967) 15–16.

4. H. M. Mead & N. Grove, *Ngā Pēpeha a ngā Tīpuna* (Wellington: Victoria University Press 2003) 323; H. W. Williams, 2003, 11.
5. P. Gloyne, personal communication, 13 July 2015.
6. www.28maoribattalion.org.nz/audio/c-company-haka; final section of haka and explanation, D. Ladelli, personal communication, 8 July 2016.
7. H. M. Mead & N. Grove, 2003, 285.
8. E. Best, *The Astronomical Knowledge of the Maori* (Wellington: Government Printer 1955) 32; L. Hakaraia, *Matariki The Māori New Year* (Auckland: Reed 2004) 29.
9. E. Best, *The Maori as he was: a brief account of Maori life as it was in pre-European days* (Wellington: Government Printer 1952) 138.
10. A. Ngata & H. M. Jones, *Ngā Moteatea: He marama rere no nga waka maha.* Vol. 2 (Wellington: Polynesian Society 2004a) 152–161.
11. Ministry for Culture and Heritage, Te Taiao – *Māori and the Natural World* (Auckland: David Bateman 2010) 22.
12. E. Best, 1952, 134–135
13. Te Wehi, 'He Wharangi Tuwhera', *Te Waka Maori o Niu Tirani* (Vol. 10, No. 9, 22 September 1874) 240.
14. T. Ruka, 'Reta ki te Etita', *Te Pipiwharauroa He Kupu Whakamarama* (No. 14, 1 April 1899) 4.
15. 'Te Aroha o Rangi-nui Kia Papatuanuku', *Te Toa Takitini* (No. 11, 1 June 1922) 10.
16. 'Ko Te Tangi A Karanga, Mo Whakatere', *Te Karere o Poneke* (Vol. 1, No. 27, 10 May 1858) 3.
17. E. Best, *Tuhoe: Children of the Mist* (Auckland: Reed 1996) 836.
18. S. Rerekura, *Puanga: Star of the Māori New Year* (Auckland: Te Whare Wānanga o Ngāpuhi-nui-tonu 2014).
19. E. Best, *Maori Agriculture* (Wellington: A. R. Shearer, Government Printer 1976) 107, 199, 215.
20. According to Wharehuia Milroy (personal communication, 20 May 2014) the saying 'whetū heri kai' can be applied when thanking people for their hospitality, especially in relation to food. Those who prepare food and serve others can be likened to the stars that bring forth the bounty of the year; E. Best, 'Maori Forest Lore', in *Transactions of the New Zealand Institute* (Vol. 42, 1909) 448.
21. J. Johansen, *Studies in Maori Rites and Myths* (Copenhagen: E. Munksgaard 1958) 17.
22. E. Best, 1976, 116.
23. T. Maxwell, *Te Kōpura*. Unpublished master's thesis, University of Waikato, New Zealand 1998, 34–35.
24. J. Binney, *Redemption Songs: A life of Te Kooti Arikirangi Te Turuki* (Wellington: Bridget Williams Books 1995) 421–422.
25. T. Maxwell, 'Ringatu planting, harvesting rites', in *Te Nupepa o te Tairawhiti* (June 2005) 22. http://www.ngaituhoe.com/files/te_ao_maori_jun05.pdf.

26. T. Maxwell, 1998, 34.
27. Ministry for Culture and Heritage, 2010, 23.
28. M. Riley, *Wise Words of the Māori: Revealing history and traditions* (Paraparaumu: Viking Sevenseas 2013) 465.
29. M. Riley, *Māori Bird Lore* (Paraparaumu: Viking Sevenseas 2001) 25.
30. E. Best, 1955, 53.
31. N. Short, *Te Kererū o Te Urewera* Unpublished master's thesis, Massey University, New Zealand 2015; Waitangi Tribunal, *Te Urewera Part VI. WAI 894* (Wellington: Waitangi Tribunal 2015) 22–24.
32. E. Best, 1996, 879.
33. P. Temara, personal communication, 15 March 2014.
34. H. M. Mead & N. Grove, 2003, 272; M. Riley, 2013, 373.
35. R. M. McDowall, *Ikawai: Freshwater fishes in Māori culture and economy* (Christchurch: Canterbury University Press 2011) 123.
36. E. Best, 1996, 813.
37. A. Walsh, 'The Cultivations and Treatment of the Kumara by the primitive Maori', in *Transactions and Proceedings of the New Zealand Institute* (Vol. 35, 1902) 17–20; W. Colenso, 'On the Vegetable Food of the Ancient New Zealanders before Cook's Visit', in *Transactions and Proceedings of the New Zealand Institute* (Vol. 13, 1880) 12; E. Best, 1952, 138.
38. H. M. Mead & N. Grove, 2003, 61.
39. 'Maori Proverbs', *Te Waka Maori o Niu Tirani* (Vol. 11, No. 13, 6 July 1875) 159.
40. H. M. Mead & N. Grove, 2003, 182, 285.
41. The term pātere is given to a form of traditional Māori song that has a fast tempo causing performers to translate the words they sing into movement.
42. A. Ngata & H. M. Jones, *Ngā mōteatea: He maramara rere no nga waka maha.* Vol. 2 (Wellington: Polynesian Society 2004a) 156–157.
43. E. Best, 1955, 53; M. Riley, 2013, 773.
44. C. Kirkwood, *Te Arikinui and the Millennium of Waikato* (Ngaruawahia: Turongo House 2001) 195; M. Riley 2013, 773.
45. E. Best, 1952, 135; J. Beattie, *Traditional Lifeways of the Southern Maori* (Dunedin: University of Otago Press in association with Otago Museum 1994) 202.
46. E. Best, 'Hau and Wairaka. The Adventures of Kupe and His Relatives', in *Journal of Polynesian Society* (Vol. 36, No. 143, 1927) 266.
47. E. Best 1955, 37.
48. H. Kerr, 'Māori Astronomy', *Paepae,* Series 1 Episode 30 (Māori Television. 18 October 2015).

MODERN MATARIKI

1. A. Anderson, J. Binny & A. Harris, *Tangata Whenua: An Illustrated History* (Wellington: Bridget Williams Books 2014) 319–349.

2. S. Myhre, 'Matariki Celebrations' *The Northland Age* 3 July 2012: http://www.nzherald.co.nz/northlandage/lifestyle/news/article.cfm?c_id=1503396&objectid=11069
3. New Zealand Government, *Te Rā o Matariki Bill/Matariki Day Bill* (Wellington: Tables Office 2009) http://www.legislation.govt.nz/bill/member/2009/0056/latest/whole.html
4. https://www.twoa.ac.nz/landing/Matariki
5. D. McGregor, *Nā Kua'āina. Living Hawai'ian Culture* (Honolulu: University of Hawai'i Press 2007) 272.
6. M. Riley, *Wise Words of the Māori: Revealing history and traditions* (Paraparaumu: Viking Sevenseas 2013) 661.

MATARIKI PROVERBS

1. 'Maori Proberbs', *Te Waka Maori o Niu Tirani* (Vol. 11, No. 13, 6 July 1875) 159.
2. H. M. Mead & N. Grove, *Ngā Pēpeha a ngā Tīpuna* (Wellington: Victoria University Press 2003) 162.
3. M. Riley, *Wise Words of the Māori: Revealing history and traditions* (Paraparaumu: Viking Sevenseas 2013) 348.
4. A. E. Brougham & A. W. Reed, *Maori Proverbs* (Wellington: Reed 1975) 124.
5. H. M. Mead & N. Grove, 2003, 181.
6. M. Riley, 2013, 348.
7. E. Best, *The Maori.* Vol. 2 (Wellington: Harry H. Tombs 1924) 489.
8. H. M. Mead & N. Grove, 2003, 182.
9. Paul Meredith, 'Matariki – Māori New Year - Heralding the New Year', *Te Ara - the Encyclopedia of New Zealand*, http://www.TeAra.govt.nz/en/matariki-maori-new-year/page-1
10. H. Kereopa, 'Matariki', *Waka Huia* (TVNZ 2001).
11. M. Riley, *Māori Bird Lore* (Paraparaumu: Viking Sevenseas 2001) 25.
12. G. Grey, *Proverbial and Popular Sayings* (London: Trubner 1857) 65.
13. A. E. Brougham & A. W. Reed, 1975, 75.
14. M. Riley, *Māori Sayings and Proverbs* (Paraparaumu: Viking Sevenseas 1990) 73-10.
15. G. Grey, 1857, 65.
16. H. M. Mead & N. Grove, 2003, 285.
17. H. W. Williams, *Dictionary of the Māori Language* (Wellington: Legislation Direct 2003) 229.
18. H. W. Williams, 2003, 385.
19. E. Best, *The Astronomical Knowledge of the Maori* (Wellington: Government Printer 1955) 53.
20. H. W. Williams, *He Whakatauki, he titotito, he pepeha* (Gisborne: Te Rau Kahikatea 1908) 28.
21. R. Te Kōkau manuscript (1898–1933).

BIBLIOGRAPHY

Allen, R. H. 1963. *Star Names, their Lore and Meaning*. New York: Dover.

Anderson, A., Binny, J., Harris, A. 2014. *Tangata Whenua: An Illustrated History*. Wellington: Bridget Williams Books.

Andrew, T. 1998. *Dictionary of Nature Myths: Legends of the Earth, Sea and Sky*. New York: Oxford University Press.

Andrews, M. 2004. *The Seven Sisters of the Pleiades: Stories from around the world*. Melbourne: Spinifex Press.

Batten, J. 2005. *Celebrating the Southern Seasons: Rituals for Aotearoa*. Auckland: Random House.

Beaglehole, E. 1938. *Ethnology of Pukapuka*. Honolulu: Bernice P. Bishop Museum.

Beattie, J. 1994. *Traditional Lifeways of the Southern Maori*. Dunedin: University of Otago Press in association with Otago Museum.

Beckwith, M. 1976. *Hawaiian Mythology*. Honolulu: University of Hawai'i Press.

Beckwith, M. 1981. *The Kumulipo: A Hawaiian Creation Chant*. Honolulu: University of Hawai'i Press.

Best, E. 1909. 'Maori Forest Lore', in *Transactions of the New Zealand Institute*. Vol. 42, 433–481.

Best, E. 1910. 'Maori Star Names', in *Journal of Polynesian Society*. Vol. 19. No. 2, 97–99.

Best, E. 1924. *The Maori*. Vol. 2. Wellington: Harry H. Tombs.

Best, E. 1927. 'Hau and Wairaka. The Adventures of Kupe and His Relatives', in *Journal of Polynesian Society*. Vol. 36, No. 143, 260–286.

Best, E. 1952. *The Maori as he was: a brief account of Maori life as it was in pre-European days*. Wellington: Government Printer.

Best, E. 1954. *Some Aspects of Maori Myth and Religion*. Wellington: Government Printer.

Best, E. 1955. *The Astronomical Knowledge of the Maori*. Wellington: Government Printer.

Best, E. 1973. *The Maori Division of Time*. Wellington: A. R. Shearer, Government Printer.

Best, E. 1976. *Maori Agriculture*. Wellington: A. R. Shearer, Government Printer.

Best, E. 1996. *Tuhoe: Children of the Mist*. Auckland: Reed.

Best, E. 2005a. *Maori Religion and Mythology*. Part 1. Wellington: Te Papa Tongarewa.

Best, E. 2005b. *Maori Religion and Mythology*. Part 2. Wellington: Te Papa Press.

Binney, J. 1995. *Redemption Songs: A life of Te Kooti Arikirangi Te Turuki*. Wellington: Bridget Williams Books.

Bogard, P. 2007. *Let There Be Night: Testimony on behalf of the dark*. Reno: University of Nevada Press.

Brady, B. 1998. *Brady's Book of Fixed Stars*. Boston: Weiser Books.

Brougham, A. E. and A. W. Reed. 1975. *Maori Proverbs*. Wellington: Reed.

Buck, P. 1938. *Ethnology of Mangareva*. Bulletin 157. Honolulu: Bernice P. Bishop Museum.

Burnham, R. 1978. *Burnham's Celestial Handbook: An Observer's Guide to the Universe Beyond the Solar System*. Vol III. New York: Dover Publications.

Clark, K. 1896. *Maori Tales & Legends*. London: David Nutt.

Colenso, W. 1880. 'On the Vegetable Food of the Ancient New Zealanders before Cook's Visit', in *Transactions and Proceedings of the New Zealand Institute*. Vol 13, 338.

Dansey, H. 1967. 'Matariki', in *Te Ao Hou: The New World*. Vol 61, 15–16.

Department of Internal Affairs. 1994. *Ngā Tāngata Taumata Rau.* Wellington: Bridget Williams Books & Department of Internal Affairs.

Dodd, E. 1967. *Polynesian Art.* New York: Dodd, Mean & Company.

Doone, T. E. 1927. *The Maori Past and Present.* London: Seeley Service and Co.

Evans, J. 1998. *The History and Practice of Ancient Astronomy.* New York: Oxford University Press.

Gifford, A. C. 1937. *In Starry Skies: The Solar System.* Wellington: New Zealand Astronomical Society.

Gopakumar, K. G. 2013. *The Great Year and Virgin Comets.* India: Gopa Kumar.

Grey, G. 1857. *Proverbial and Popular Sayings.* London: Trubner.

Grey, G. 2010. *Ko nga Moteatea me nga Hakirara o nga Maori.* Wellington: Stokes.

Hail, R. 2000. *Cherokee Astrology. Animal Medicine in the Stars.* Rochester: Bear & Company.

Hakaraia, L. 2004. *Matariki: The Māori New Year.* Auckland: Reed.

Handy, E. S. 1923. *The Native Culture in the Marquesas.* Bulletin 48. Honolulu: Bernice P. Bishop Museum.

Handy, E. S. C. 1927. *Polynesian Religion.* Honolulu: Bernice P. Bishop Museum.

Hard, R. 2004. *The Routledge Handbook of Greek Mythology.* New York: Routledge.

Hard, R. 2015. *Eratosthenes and Hyginus Constellation myths with Aratus's Phaenonena.* Oxford: Oxford University Press.

Harness, D. 2000. *The Nakshatras. The Lunar Mansions of Vedic Astrology.* Delhi: Motilal Banarsidass.

Harris, H., Matamua, R., Smith, T., Kerr, H., Waaka, T. 2013. 'A review of Māori Astronomy in Aotearoa-New Zealand', in *Journal of Astronomical History and Heritage.* Vol. 16, No. 3, 325–336.

Henry, T. 1907. 'Tahitian Astronomy', in *Journal of Polynesian Society.* Vol. 16. No. 2, 101–104.

Hetherington, E. W. & N. S. 2009. *Astronomy and Culture.* Santa Barbara: Greenwood Press.

Highland, G., Force, R., Howard, A., Kelly, M., Sinoto, Y. 1967. *Polynesian Culture History*. Essays in Honor of Kenneth P. Emory. Honolulu: Bernice P. Bishop Museum Press.

'Honorific Terms, Sacerdotal Expressions, Personifications, etc., Met with in Maori Narrative', in *Journal of the Polynesian Society*. 1926. Vol. 35, No. 140, 333–334.

Horowitz, W. 1998. *Mesopotamian Cosmic Geography*. Indiana: Eisenbrauns.

Howard, A. 1971. *Polynesia: Readings on a Culture Area*. San Francisco: Chandler.

Howe, K. R. 2006. *Vaka Moana: Voyages of the Ancestors. The Discovery and Settlement of the Pacific*. Auckland: David Bateman.

Hughes, A. 2016. *Matariki: Everything there is to know about it*. 6 June 2016. http://www.stuff.co.nz/science/80739387/Matariki-Everything-there-is-to-know-about-it

Johansen, J. 1958. *Studies in Maori Rites and Myths*. Copenhagen: E. Munksgaard.

Johns, A. 2014. *Baba Yaga – The Ambiguous Mother and Witch of the Russian Folktale*. New York: Peter Lang.

Johnson, D. 2006. *Night Skies of Aboriginal Australia*. Sydney: Sydney University Press.

Johnson, R., Mahelona, J., Ruggles, C. 2015. *Nā Inoa Hōkū. Hawaiian and Pacific Star Names*. Bognor Regis: Ocarina Books.

Jones, L. V. 2010. *Stars and Galaxies*. Santa Barbara: Greenwood Press.

Joseph, F. 2005. *The Atlantis Encyclopedia*. New Jersey: New Page Books.

Kereopa, H. 2001. 'Matariki', *Waka Huia*. TVNZ.

Kerr, H. 2015. 'Māori Astronomy', *Paepae*. Māori Television. 18 October 2015.

Kien, J. 2003. *The Battle Between the Moon and Sun*. Boca Raton: Universal Publishers.

Kirkwood, C. 1999. *Koroki, my King*. Ngaruawahia: Turongo House.

Kirkwood, C. 2001. *Te Arikinui and the Millennium of Waikato*. Ngaruawahia: Turongo House.

'Ko Te Tangi A Karanga, Mo Whakatere'. 1858. *Te Karere o Poneke*. 10 May 1858. Vol. 1, No. 27, 3.

Krupp, E. 1991. *Beyond the Blue Horizon: Myths and Legends of the Sun, Moon, Stars and Planets.* New York: HarperCollins.

Leaman, T. M., Hamacher, D. W., Carter, M. T. 2011. 'Aboriginal Astronomical Traditions from Ooldea, South Australia, Part 2: Animals in the Ooldean Sky', in *Journal of Astronomical History and Heritage*. Vol. 19, No. 1, Preprint.

Leather, K. & Hall, R. 2004. *Tātai Arorangi: Māori Astronomy. Work of the Gods.* Paraparaumu: Viking Sevenseas.

Levin, W., Reeve, R., Salmoiraghi, F., Ulrich, D. 1995. *Kaho'olawe, Nā Leo o Kanaloa. Chants and stories of Kaho'olawe.* Honolulu: Ai Pōhaku Press.

Llewellyn. 2012. *Llewellyn's 2013 Sabbats Almanac. Samhain 2012 to Mabon 2013.* Woodburry: Llewellyn Worldwide Ltd.

McDowall, R. M. 2011. *Ikawai: Freshwater fishes in Māori culture and economy.* Christchurch: Canterbury University Press.

McGregor, D. 2007. *Nā Kua'āina. Living Hawaiian Culture.* Honolulu: University of Hawai'i Press.

Magli, G. 2005. *Mysteries and Discoveries of Archaeoastronomy. From Giza to Easter Island.* New York: Praxis Publishing.

Magli, G. 2016. *Archaeoastronomy. Introduction to the Science of Stars and Stones.* Cham: Springer.

Mahupuku, H. 1854. *Whakapapa Tuupuna.* MS, private collection, Whakatāne. Cited by P. Hohepa in his translation document (1992).

Makemson, M. 1941. *The Morning Star Rises: An Account of Polynesian Astronomy.* New Haven: Yale University Press.

'Maori Proverbs'. 1875. *Te Waka Maori o Niu Tirani.* 6 July 1875. Vol. 11, No. 13, 159.

Maryboy, N. & Begay, D. 2010. *Sharing The Skies: Navajo Astronomy.* Tucson: Rio Nuevo Publishers.

Maxwell, T. 1998. *Te Kōpura.* Unpublished master's thesis. University of Waikato, New Zealand.

Maxwell, T. 2005. 'Ringatu planting, harvesting rites', in *Te Nupepa o te Tairawhiti*. June, 22–23. Retrieved from http://www.ngaituhoe.com/files/te_ao_maori_jun05.pdf

Mead, H. 1981. *Nga Taonga Tuku Iho a Ngati Awa: Ko nga tuhituhi a Hamiora Pio, Te Teko (1885–1887)*. Wellington: Department of Māori, Victoria University of Wellington.

Mead, H. M. 2003. *Tikanga Māori: Living by Māori Values*. Wellington: Huia.

Mead, H. M., & Grove, N. 2003. *Ngā Pēpeha a ngā Tīpuna*. Wellington: Victoria University Press.

Ministry for Culture and Heritage. 2010. *Te Taiao – Māori and the Natural World*. Auckland: David Bateman.

Mitcalfe, B. 1961. 'Te Rerenga Wairua – Leaping Place of the Spirits', in *Te Ao Hou: The New World*. No. 35 (June 1961), 38–42.

Moyle, R. 2011. *Takuu Grammar and Dictionary: A Polynesian language of the South Pacific*. Canberra: The Australian National University.

Myhre, S. 2012. *The Northland Age* 3 July 2012. 'Matariki Celebrations', http://www.nzherald.co.nz/northlandage/lifestyle/news/article.cfm?c_id=1503396&objectid=11069213

Ngata, A. & Jones, H. M. 2004a. *Ngā Mōteatea: He maramara rere no nga waka maha*. Vol. 2. Wellington: Polynesian Society.

Ngata, A., & Jones, P. T. H. 2004b. *Ngā Mōteatea: He maramara rere no nga waka maha*. Vol. 3. Wellington: Polynesian Society.

Ngata, A., & Mead, H. M. 2004. *Ngā Mōteatea: He maramara rere no nga waka maha*. Vol. 4. Wellington: Polynesian Society.

New Zealand Government. 2009. *Te Rā o Matariki Bill/Matariki Day Bill*. Wellington: Tables Office. http://www.legislation.govt.nz/bill/member/2009/0056/latest/whole.html

Olcott, W. T. 2004. *Star Lore: Myths, Legends and Facts*. New York: Dover Publications.

Orbell, M. R., & Moon, G. 1985. *The Natural World of the Maori*. Auckland: David Bateman.

Radio New Zealand. 1981. *Whaikoorero: Ceremonial Farewells to the Dead.* Wellington: Continuing Education Unit, Radio New Zealand.

Rappengluck, M. 1997. 'The Pleiades in the "Salle des Taureaux" grotte de Lascaux. Does a rock picture in the cave of Lascaux show the open star cluster of the Pleiades at the Magdalénien era ca 15.300 BC?', in *IVth SEAC Meeting 'Astronomy and Culture'*, 217–225.

Red Star, N. 2000. *Star Ancestors. Extraterrestrial Contact in the Native American Tradition.* Rochester: Bear & Company.

Reeves, W. 1898. *The Long White Cloud Aotearoa.* London: Horace Marshall & Son.

Renshaw, S. & Ihara, S. A, 'Cultural History of Astronomy in Japan', in Selin, H. 2000. *Astronomy Across Cultures: The history of non-western astronomy.* Berlin: Springer Science+Business Media.

Rerekura, S. 2014. *Puanga: Star of the Māori New Year.* Auckland: Te Whare Wānanga o Ngāpuhi-nui-tonu.

Riley, M. 1990. *Māori Sayings and Proverbs.* Paraparaumu: Viking Sevenseas.

Riley, M. 2001. *Māori Bird Lore.* Paraparaumu: Viking Sevenseas.

Riley, M. 2013. *Wise Words of the Māori: Revealing history and traditions.* Paraparaumu: Viking Sevenseas.

Roberts, M., Weko, F., Clarke, L. 2006. *Maramataka: the Māori Moon Calendar.* Research Report No. 283. Christchurch: Lincoln University.

Rolleston-Cummins, T. 2008. *The Seven Stars of Matariki.* Wellington: Huia.

Ruggles, C. 2005. *Ancient Astronomy: An Encyclopedia of Cosmologies and Myth.* Santa Barbara: ABC-CLIO.

Ruka, T. 1899. 'Reta ki te Etita', in *Te Pipiwharauroa He Kupu Whakamarama.* No. 14 (1 April 1899), 3–4.

Short, N. 2015. *Te Kererū o Te Urewera.* Unpublished master's thesis. Massey University, New Zealand.

Sissons, J. 2014. *The Polynesian Iconoclasm. Religious Revolution and the Seasonality of Power.* New York: Berghahn Books.

Smith, P. 1900. 'Wars of the northern against the southern tribes of New Zealand in the nineteenth century Part III', in *Journal of the Polynesian Society*. Vol. 9, No. 1, 1–37.

Smith, P. 1978. *The Lore of the Whare Wānanga*. New York: AMS Press.

Steele, P. R. 2004. *Handbook of Inca Mythology*. Santa Barbara: ABC-CLIO.

Taonui, A. 1875. 'Ki Te Kai Ta o Te Wananga' in *Te Wananga*. 21 August 1875. Vol. 2, No. 16, 163.

Tawhai, W. 2009. 'Te Maramataka Māori', *Waka Huia*. TVNZ, 13 June 2009.

Tāwhai, W. 2013. *Living by the Moon: Te Maramataka a Te Whānau*-ā-*Apanui*. Wellington: Huia.

'Te Aroha o Rangi-nui Kia Papatuanuku'. 1922. *Te Toa Takitini*. 1 May 1922, No. 10, 13.

'Te Aroha o Rangi-nui Kia Papatuanuku'. 1922. *Te Toa Takitini*. 1 June 1922, No. 11, 10.

Te Kōkau, R. 1898–1933. Unpublished manuscript, private collection, Ruatāhuna.

Te Rangi, T. H. 1860. 'Tera Matariki' in *The Maori Messenger – Te Karere Maori*. 30 November 1860, Vol. 7, No. 18, 39–40.

Te Taura Whiri i te Reo Māori. 2005. *Matariki: Aotearoa Pacific New Year*. Wellington: Te Taura Whiri i te Reo Māori.

Te Taura Whiri i te Reo Māori. 2010. *Matariki*. Wellington: Te Taura Whiri i te Reo Māori.

Te Wehi. 1874. 'He Wharangi Tuwhera', in *Te Waka Maori o Niu Tirani*. 22 September 1874, Vol. 10, No. 9, 239–240.

Te Whetu o Te Tau. Akarana, 1 Hune 1858, Vol. 1, No. 1. Retrieved from: https://paperspast.natlib.govt.nz/newspapers/WHETU18580601.2.5?query=Te%20Whetu%20o%20te%20tau.%20Akarana.

Tobin, J. 2002. *Stories from the Marshall Island*. Honolulu: University of Hawai'i Press.

Tohunga. 1935, October 1. The Wisdom of the Maori. *The New Zealand Railway Magazine*. Vol. 10, No. 7, 45.

Tregear, E. 1904. *The Maori Race*. Wanganui: A. D. Willis Printer and Publisher.

Tuahine, H. 2015. *Te Tāhū o Ranginui: Whakatūria te Whare Kōkōrangi*. Unpublished master's thesis. University of Waikato, New Zealand.

Turner, G. 1884. *Samoa, A Hundred Years Ago and Long Before*. London: Macmillan.

Turongo House. 1983. *Te Tumu Korero*. Ngaruawahia: The House.

Tyerman, D. & Bennet, G. 1841. *Voyages and Travels Round the World*. London: John Snow.

Waitangi Tribunal. 2015. *Te Urewera Part VI. WAI 894*. Wellington: Waitangi Tribunal.

Walsh, A. 1902. 'The Cultivations and Treatment of the Kumara by the primitive Maori', in *Transactions and Proceedings of the New Zealand Institute*. Vol 35, 12–24.

Webb, S. 2001. *Measuring the Universe. The Cosmological Distance Ladder*. Leicestershire: Springer-Praxis.

White, J. 1887. *Ancient History of the Maori, His Mythology and Traditions Horouta or Takitumu Migration*. Vol. I. Wellington: Government Printer.

White, J. 1888a. *Ancient History of the Maori, His Mythology and Traditions Tai-Nui*. Vol. IV. Wellington: Government Printer.

White, J. 1888b. *Ancient History of the Maori, His Mythology and Traditions Tai-Nui*. Vol. V. Wellington: Government Printer.

Williams, H. W. 1908. *He Whakatauki, he titotito, he pepeha*. Gisborne: Te Rau Kahikatea.

Williams, H. W. 2003. *Dictionary of the Māori Language*. Wellington: Legislation Direct.

Wood, F. & K. 2014. *Homer's Secret Odyssey*. London: The History Press.

INDEX

Page numbers in italic indicate images.